Experiments in Cell and Molecular Biology

A Laboratory Manual for Undergraduate Instruction or Concepts and Techniques for the Undergraduate Laboratory

RIVER PUBLISHERS SERIES IN LABORATORY MANUALS FOR LIFE SCIENCES

Series Editor:

TRACY S. TRAN
Rutgers University
USA

The scope and aim of this series is to provide diverse sets of laboratory experience in the form of biological experiments (activities) for college level students ranging from introductory, intermediate to advance levels. This series will have, approximately, 8 laboratory manuals divided into Part 1 (4 manuals) and Part II (4 manuals) covering a diverse subfield of biological sciences, including but not limited to:

- Cell & Molecular Biology
- Biochemistry
- Genetics
- Plant Science
- Microbiology
- Neuroanatomy
- Immunology
- Ecology

For a list of other books in this series, visit www.riverpublishers.com

Experiments in Cell and Molecular Biology

A Laboratory Manual for Undergraduate Instruction or Concepts and Techniques for the Undergraduate Laboratory

Miguel Cervantes-Cervantes

Rutgers, The State University of New Jersey, Newark, NJ, USA

River Publishers

Routledge
Taylor & Francis Group
NEW YORK AND LONDON

Published 2026 by River Publishers
River Publishers
Broagervej 10, 9260 Gistrup, Denmark
www.riverpublishers.com

Distributed exclusively by Routledge
605 Third Avenue, New York, NY 10017, USA
4 Park Square, Milton Park, Abingdon, Oxon OX14 4RN

Experiments in Cell and Molecular Biology / by Miguel Cervantes-Cervantes.

Routledge is an imprint of the Taylor & Francis Group, an informa business

ISBN 978-87-4381-320-0 (paperback)
ISBN 978-87-4381-323-1 (online)
ISBN 978-87-4381-322-4 (master ebook)
ISBN 978-87-4381-325-5 (ePub)

DOI: 10.1201/9788743813224

Cover Image

Transmission electron microscope image of an aleurone cell from wheat (*Triticum aestivum* INIA 66R). Aleurone layers were treated with gibberellic acid for 48 h after imbibition. Programmed cell death is evident from the cell shrinkage. membrane blebbing, chromatin condensation, and the cytoplasm disorganization. Notice the seemingly healthy cell to the right. Size bar = 5 μm. Electron micrograph by M. Chávez Flores.

Contents

Foreword

The remarkable advances in cell and molecular biology over the past several decades have transformed our understanding of life at its most fundamental level. From elucidating the molecular basis of heredity to characterizing complex cellular signaling networks, modern biology has become increasingly dependent on experimental approaches that integrate biochemical, genetic, computational, and molecular techniques. Consequently, undergraduate education in the life sciences must provide students not only with a sound theoretical framework but also with meaningful laboratory experiences that cultivate scientific reasoning and technical competence.

Experiments in Cell and Molecular Biology is a laboratory manual designed for undergraduate instruction, covering concepts and techniques for undergraduate students, and represents a thoughtful and timely contribution to undergraduate science education. The manual has been designed to guide students beyond the passive acquisition of knowledge and into the active practice of scientific investigation. Through a sequence of carefully organized laboratory exercises, students are introduced to both the conceptual foundations and experimental methodologies that define contemporary cell and molecular biology.

The ten exercises presented in this volume are distinguished by their logical progression and pedagogical coherence. Beginning with introductory topics such as bioinformatics and the preparation of biological buffers, students acquire essential laboratory skills while developing an appreciation for the quantitative and analytical nature of biological inquiry. As the course advances, students encounter increasingly sophisticated techniques, including protein quantification through the Bradford assay, enzyme kinetics, polyacrylamide gel electrophoresis, molecular evolution, genetic fingerprinting, and signal transduction analysis. This graduated approach enables learners to build confidence while mastering methods that are central to modern biological research.

Particularly commendable is the emphasis placed on experimental design, data interpretation, and hypothesis-driven investigation. Rather than presenting laboratory exercises as isolated procedures, the manual encourages students to engage with the scientific process itself. Such an approach reflects the realities of contemporary research, where critical thinking, quantitative analysis, and the ability to draw evidence-based conclusions are as important as technical proficiency.

The biological themes explored throughout the manual encompass many of the central pillars of modern cell biology, including cellular composition, subcellular organization, metabolism, membrane function, intracellular communication, and the molecular mechanisms governing gene expression and regulation. By examining these topics through direct experimentation, students gain a deeper appreciation for the dynamic and interconnected nature of living systems. Equally important, they develop an understanding of how biological knowledge is generated through observation, measurement, analysis, and interpretation.

As educators, we recognize that the laboratory occupies a unique and indispensable role in science instruction. It is within the laboratory that abstract concepts acquire tangible meaning and where students learn to navigate the uncertainties, challenges, and rewards inherent in scientific discovery. Manuals such as this one play a critical role in fostering these experiences by providing structured opportunities for inquiry while encouraging intellectual independence and curiosity.

This volume will serve as a valuable resource for undergraduate students, instructors, and laboratory coordinators alike. Whether used as a companion to a lecture course or as the foundation of a stand-alone laboratory curriculum, it provides a rigorous and engaging introduction to the principles and practices of cell and molecular biology. Students who complete these exercises will emerge with a stronger conceptual understanding of biological systems and with practical skills that will serve them well in advanced coursework, research laboratories, and future professional endeavors.

The author is to be commended for creating a manual that successfully integrates foundational knowledge, experimental methodology, and scientific inquiry. In doing so, he has produced a resource that not only teaches techniques but also cultivates the habits of mind essential to the biological sciences. It is my pleasure to recommend this work to students and instructors seeking a meaningful and modern introduction to the experimental study of the cell.

Tracy S. Tran, Ph.D.
Professor and Director of Undergraduate Program in Biology
Rutgers University

Description

This laboratory manual is designed for students of biological sciences focusing on the following areas: cell chemical components, cell structure, metabolism and signal transduction. This laboratory manual provides practical experiences beyond demonstrations, as well as hands-on activities and guided experimental design with a focus on cell and molecular biology techniques. There are 10 laboratory exercises, with increasing depth of knowledge and difficulty level. While the experimental exercises in this laboratory manual are best complemented by a lecture course at a similar level that covers foundational knowledge in cell and molecular biology, it can be taught as a stand-alone biology laboratory course for sophomore level and above.

Prerequisites

While the materials covered in this laboratory manual are at the level of a college mid-level Biology course, the students should have some knowledge of the fundamental Concepts in Biology, and General Chemistry. Additional foundational knowledge in the topics of cell and molecular biology, particularly in energy utilization and transfer, protein and nucleic acid structure and function, transcription, translation, and genetic regulation will be helpful.

Mission Statement

The purpose of this laboratory manual is to provide students with factual and conceptual tools in the field of cell and molecular biology, thereby facilitating the pursuit of higher-level coursework in biological and life sciences-related fields. This is a beginner-to-mid-level course for students with a general background in physics and chemistry interested in biology as a major or minor. This Experiments in Cell and Molecular Biology laboratory manual aims to introduce students to procedural knowledge through the study and analysis of intracellular processes, with an emphasis on eukaryotic models, as well as the application of specific methods in cell and molecular biology. The theoretical component of the course may be delivered through a parallel lecture course at the same or similar level and focus, or in the form of brief stand-alone reviews at the beginning of each laboratory session.

Learning Goals

The course aims to provide students studying biological sciences with a deep understanding of basic phenomena in cell and molecular biology, in preparation for higher-level coursework. Topics covered are: The chemical components of the cell; subcellular structure and methods of study; thermodynamics and metabolism; membrane biology, energy utilization and transfer; protein and nucleic acid structure and function; transcription, translation, and genetic regulation.

Learning Objectives

Through selected readings, lectures, discussions, and occasional group activities, students are encouraged to learn on their own about the main processes taking place in the cell from a molecular perspective. After successfully completing the experiments in this laboratory course, students will be able to

- ✓ Describe the general structure of biomolecules as well as their role in cellular metabolism and the flow of genetic information.
- ✓ Apply information and concepts to bioenergetics and the use of energy by cell.

- ✓ Acquire information on the principles of membrane transport mechanisms and their role in important physiological processes at the organismal level.
- ✓ Acquire concepts and general principles on gene expression and its regulation.
- ✓ Review their knowledge on the concepts and general principles of eukaryotic signal transduction.
- ✓ Develop the skills to read, interpret and apply general information in the fields of cell and molecular biology.
- ✓ Evaluate contemporary hypotheses on the functional mechanisms of the cell.
- ✓ Reinterpret and postulate alternative hypotheses or ideas to explain or describe the phenomena studied in the course.

General Laboratory Instructions

(A) Laboratory Safety Rules

The following safety rules apply to this laboratory course and must be reviewed and always followed. It is not possible for instructors to monitor every action; therefore, you are expected to take personal responsibility for your safety and the safety of others.

1. Eye protection must be always worn in the laboratory. Extra caution should be taken if you wear contact lenses.
2. Food and drink are strictly prohibited in the laboratory. Do not bring food or beverages into the lab, even if you intend to consume them later.
3. Do not dispose of food wrappers or containers in laboratory trash receptacles. Safety inspectors will not distinguish between food consumed inside or outside the laboratory.
4. Never pipet by mouth. Always use pipette pumps. For small volumes (less than 1 mL), use mechanical pipettors with disposable plastic tips.
5. Long hair must be tied back to prevent contact with flames, chemicals, or equipment moving parts.
6. Keep laboratory benches clean and free of unnecessary items.
7. Store personal belongings if lockers are provided for your class during laboratory sessions in which case, you must bring your own lock.
8. When working with volatile or corrosive substances, use the fume hoods. Do not remove chemicals from the fume hood area.
9. Clean up spills immediately. If you are unsure how to proceed, ask for immediate assistance.
10. Only wear closed-toe shoes at all times in the laboratory. Sandals and open-toe footwear are not permitted.
11. Dispose of broken glass only in containers labeled "Broken Glass," not in biohazard containers.
12. Remain alert and behave in a professional manner at all times. Horseplay is strictly prohibited in the laboratory.

This is not an all-encompassing list of safety practices. You should consult the Laboratory Safety Rules notice elaborated by the department that offers this lab.
Wearing a laboratory coat while performing experiments is mandatory. No toxic or radioactive substances are used, but dyes or staining solutions may get onto your clothes or on your skin.
Ask your instructors about stores, including websites where you can purchase lab coats. Some vendors can be found online.
Purchasing a lab coat online may bring some challenges other than choosing the right size. For example, you want a coat with a sturdy fabric, impermeable to mild organic solvents and fire-resistant.

(B) Procedures for Emergency Medical Care

1. Life-threatening Emergencies

   The number to call in case of a medical (or other emergencies) depends on the country in which you are located. Call 911 in the United States and other countries in the American continent; 999 or 112 in the United Kingdom; 112 in the European Union, Asia and Africa; etc.
2. Non-lethal Medical Emergencies and Accidents.

(a) Notify your campus police or the student health service; state the problem, and the patient's condition, name and location.
(b) The campus physician or nurse will evaluate the case and render any required care or indicate that the patient should be taken to the nearest hospital or clinic.

3. Chemical Hygiene

   Many chemical compounds are toxic, flammable, explosive or corrosive. It is not possible to remove every hazardous material from student experiments, nor even desirable, since this would prevent the students from learning how to deal with routine hazards, an important part of laboratory training.
4. Every effort is made in the lab to minimize certain hazards and to make students aware of any others and give supervision. However, for many compounds so little is known about toxicity and mutagenicity towards developing embryos or the unborn that it is recommended that any woman who knows or suspects she is pregnant, or who is trying to become pregnant, postpone taking this course.

(C) Guidelines for Keeping Laboratory Notebooks

Good laboratory notebook-keeping is not only part of the skills that you must acquire in this experimental laboratory course, it is also the way to document your activities, results, and thoughts related to your experiments. When you eventually work in the laboratory, the clinic, or the field, you must keep track of your data as official records, which are important during corroboration or repetition of your results or for patent purposes.
Please get a bound laboratory notebook. A good notebook to be used in a science laboratory (for instruction or research) must be bound, have pages that are numbered, preferably printed with a quadrille pattern (no lines or blank pages), slightly smaller than (or up to) letter size (8.5 × 11 in, i.e. 28 × 21 cm). Here are some suggested brands:

- National Brand Quad Ruled Computation & Lab Notebook ("Rediform") 43-591 or 43-648 (in the United States, this is also Staples Item 567644; other stores, such as Target, also sell this notebook)
- Ampad 22-156 or 2-157
- Roaring Spring 77648
- Tops Business Forms Lab & Research Book 35061

The first thing to do is to number the pages of your book with **ink** if they are not already numbered (all of the brands above satisfy these requirements). Always use ink to write on your notebook.
Second, you should set aside the *first two sheets for a table of contents*. Notebooks will be checked every laboratory session with your "pre-lab," as indicated by the instructor. You should read the entire laboratory exercise for the session so you have a clear idea of what you are going to do. Then, write or print and cut and paste the lab up to the procedure.
A brief note about writing the different sections of your lab as you go through the experiment: Pages of lab notebooks are asymmetrical, that is, the odd-numbered pages should contain the description of the procedure to follow (you may paste a print-out of the protocol). Even-numbered pages should display the description of composition of solutions (if any were prepared). This laboratory manual provides special pages for your personal notes and observations and the results as graphs, pictures, diagrams, descriptions, etc. Your instructor will indicate if they prefer to use the pages of the notebook continuously instead.
The format for keeping the notebook must include the sections below.

Title of the experiment(s)

Introduction: This part should be short, one or two paragraphs. Please *do not paraphrase the introductions to the labs in the manual and do not copy from textbooks or journals verbatim*; instead, come up with your own.

Date: Preferably in the "international" system (day/month/year, as in **8Nov26**). It is fine if you keep the American style (month/day/year, as in 11/08/26), as long as you use it consistently throughout your entire notebook.

Objective: A very short statement of the purpose(s) of the experiment.

Procedure: Refers to the protocol(s) to be followed to perform the experiment and can be given as a flow chart, a bulleted or numbered list, or a table. Please *write* this in your notebook; *no pasting of photocopies or printouts will be allowed*, except for graphs produced in Excel, pictures or drawings resulting from your experimental results or as indicated by your instructor. If you use pasted materials, the score calculations will vary from school to school. *Do not make a list of materials and equipment*. Such a list is *unnecessary* for your lab notebook entry, as materials, equipment and reagents are already written up in the procedure.

Records: Of the procedures and the data collected during the session. Data are the numbers, graphs, pictures, illustrations, etc. that you recorded as you conducted the experiment. The way data are recorded will depend on the laboratory, but it may include readings from the spectrophotometer, or the pH-meter, descriptions, drawings, tables, graphs, scans, photographs, etc. Data are just the facts without any interpretation of what they mean. These *must* be labeled with a number and a descriptive title (e.g. Figure 5.1. *p*-Nitrophenol standard curve). Always label both axes on all graphs and all columns of a table, including any units of measurement if applicable (e.g. reaction time (s); A_{600}; reaction rate (μmol/min)).

Results, analysis and conclusions: These should be included in your notebook.

Discussion: This section should contain an evaluation of the significance of the results. Try to bring everything together here. Point out why you think the observed results were obtained. Remember: "no result" is a result. Any opinions stated in your discussion must have a sound basis on fact. Mention in your discussion if any calculation helped you in particular in interpreting your data.
Discuss and interpret the information you obtained in the experiment. Indicate if the data were what you expected, according with the purpose of the lab. Refer to your tables or figures (e.g., "As it can be seen in Figure 2, two distinct bands are present and correspond to 100 and 400 bp"). If you obtain unexpected results or made any mistakes, this is the right place to include them; learning from your own errors is an important part of the learning process (e.g. "although we expected a pH of 7.5, the pH-meter indicated 6.8" or "I neglected to add 0.25 -mL aliquots and added the base in increments of 0.5 mL instead. As a result, I missed the midpoint in the titration procedure."). You may include suggestions on how to improve your results.

Bibliography: Please cite references correctly. A suggested citing style guide can be found on the next pages. You may use the examples below. Styles vary from journal to journal, but you may stick to a standard form that feels comfortable. You may choose not to follow this citation in which case you must choose a style of your own and follow it.
Please do not use this manual as a bibliographic source. The writings here are based on other people's work and have been adapted by the author. In case of plagiarism, the penalty goes from assigning a grade of 0 (zero) for the entire lab to a full report to an Academic Integrity Facilitator and the subsequent proceedings.

(D) Guidelines for the Preparation of Laboratory Reports

Please notice that the style for writing a report is different from notebook keeping. The number of points may vary depending on the grading criteria and rubrics established by the instructors.

i. **Title Page**

Include the title of the experiment, your name, your instructor's name, your group number, and the date the lab report was submitted. The title says what you did. It should be brief (aim for ten words or less) and describe the main point of the experiment or investigation. Begin your title using a keyword instead of an article like "The" or "A".

ii. **Introduction**

The introduction should be about one page and should provide some background knowledge of the lab and explain its objective or purpose. You must state the purpose of the experiment and the hypothesis being tested. The introduction should *not be copied from the lab manual*; you must write your own.

iii. **Methods**

Describe the approach you used to complete your investigation. *Do not* list the specific steps you followed or the specific materials you used, as this can be found in the lab manual. Instead, summarize what you did in one or two paragraphs.

iv. **Data/Results**

Data are the numbers, graphs, pictures, illustrations, etc. that you recorded as you conducted the experiment. Data are just the facts without any interpretation of what they mean. They are presented in the form of tables or graphs. These *must* be labeled with a number and a descriptive title (e.g. Figure 1. Titration curve of glycine with NaOH). Always label both axes on all graphs and all columns of a table, including any units of measurement if applicable (e.g. Reaction time (s); A_{600}; Reaction rate (µmol/min).

v. **Discussion**

Discuss your results, including any calculations that help to analyze the data. As you go through the data, interpret and discuss. Say whether or not the data are consistent with what you would expect if your hypothesis were true. In the text of your report, refer to tables and graphs by number (As you can see in Table 1). This is also where you would mention and discuss any mistakes you might have made while conducting the investigation. You may also describe ways the study might have been improved.

vi. **Conclusions**

The conclusion is typically a single paragraph that sums up what happened in the experiment. If applicable, whether your hypothesis was accepted or rejected, and what this means.

vii. **References**

Any outside sources that were used in the writing of the report need to be *cited within the text of the report and listed in the reference section*. Instructions for properly citing and referencing outside sources can be found in the next section. Remember, do *not* reference the lab manual.

viii. **Grammar/format**

Write several drafts of your lab report and edit them for correct grammar before submission, including spelling, complete and structured sentences, punctuation, etc.

(E) Citation Style Guide

Composing a bibliography comprises two aspects: Citing within the text and putting down the references cited.

In-text Citing

When citing as you write, there is a difference if the referenced work was produced by one or more authors. For example:

One author: (Nobel, 2003)

Two authors: (Schroeder and Hagiwara, 1990)

Three or more authors: (Bowsher *et al.*, 2008)

When citing in the text, do not write the authors' initials; only their last name should be written and respecting as much as possible the original spelling, *e.g.* (Lütcke, 1987), is preferred over (Lutcke, 1987). Regarding the locution *et al.*, it comes from the Latin *et alli* ("and friends"), so there should not be a period after the word *et* and no plural *s* after *al.* Also, the requirement for adding (or not) a comma between the author names and the year varies from one citation style to another.

If you are describing work done by several people, both the name of the author(s) and the year should be in parenthesis: "Measurements of tendril tension were made daily (Matista and Silk, 1997)." If quoting or mentioning the name of the authors, it is correct to only write the year in parenthesis: "Nobel (2003) described an equation for K^+ fluxes in guard cells." Standard abbreviations are preferred when citing journals. You may omit the issue number, but the volume and the first and final page of the article should be indicated; the volume number should be either in bold type or underlined (see examples below).

You should not cite references that you did not consult, even if they are citations in articles that you have read or at the end of the lab exercises in this manual. It is unethical to cite works that you did not read.

How to cite different types of publications in the Bibliography

Book

Alberts A, Hopkin K, Johnson A, Morgan D, Roberts K, Walter P, Heald R (2023) *Essential Cell Biology*, 6th edition. W. W. Norton, New York. Pp. 554-577.

Biswal UC, Biswal B, Raval MK (2003) *Chloroplast Biogenesis: From Proplastid to Gerontoplast.* Kluwer Academic Publishers, Dordrecht, 353 pp. [NB: This is not a citation from a portion of a publication, but the reference is to the book as a whole; hence, the abbreviation pp. goes after the number of pages.]

Tansey JT (2019) *Biochemistry: An Integrative Approach,* 1st edition. John Wiley & Sons, New York. Pp. 790-820.

Review chapter in a book

Byron K, Cervantes-Cervantes M, Wang JTL (2010) Biological Informatics: Data, Tools, and Applications. Pp. 59-69 in: Maulik U, Bandyopadhyay S, Wang JTL (eds.), *Computational Intelligence and Pattern Analysis in Biological informatics.* John Wiley & Sons, New York.

Schroeder JI, Hagiwara S (1990) Voltage-dependent activation of Ca^{2+}-regulated anion channels and K^{+} uptake channels in *Vicia faba* guard cells. Pp. 144-150, in: Leonard RT and Hepler PK (eds.), *Calcium in Plant Growth and Development, Current Topics in Plant Physiology, Vol. 4.* American Society of Plant Physiologists, Rockville, Maryland.

Robertson, D., I. Anderson, and M. Bachmann. 1978. Pigment-deficient mutants: Genetic, biochemical and developmental studies. Pp. 461-494, in: Walden, D. (ed.), *Maize Breeding and Genetics.* John Wiley & Sons, New York.

Journal article

Abduallah Y, Turki T, Byron K, Du Z, Cervantes-Cervantes M, Wang JT (2017) MapReduce Algorithms for Inferring Gene Regulatory Networks from Time-Series Microarray Data Using an Information-Theoretic Approach. Biomed Res Int 2017:6261802

Ngoua Meye Misso RL, Nsole Biteghe FA, Obiang CS, Ondo JP, Gao N, Cervantes-Cervantes M, Vignaux G, Vergeade A, Engohang-Ndong J, Mendene HE, Mabika B, Abessolo FO, Obame Engonga LC, De La Croix-Ndong J (2020) Effect of aqueous extracts of *Ficus vogeliana* Miq and *Tieghemella africana* Pierre in 7,12-Dimethylbenz(*a*)anthracene-induced skin cancer in rats. J Ethnopharmacol **263**:113244.

Szabo CM, Matsumura Y, Fukura S, Martin MB, Sanders JM, Sengupta S, Cieslak JA, Loftus TC, Lea CR, Lee HJ, Koohang A, Coates RM, Sagami H, Oldfield E (2002) Inhibition of geranylgeranyl diphosphate synthase by bisphosphonates and diphosphates: A potential route to new bone antiresorption and antiparasitic agents. J Med Chem **45**:2185-2196.

Online-only journal articles

When searching the internet for *bona fide*, authentic scientific journal articles, you may find three different types of journals:

(A) *Exclusively online journals*, i.e., those which were established in electronic format. For example, the several journals published under the umbrella of PLOS, the Public Library of Science (www.plos.org/publications/journals/).
(B) *Journals that are transitioning from printed form to electronic.* It has been reported that the many journals and magazines published by the American Chemical Society are currently available only through the web (chronicle.com/blogs/wiredcampus/chemistry-journals-go-digital-only/7264).
(C) *Printed journals* (usually available in the library or in a professor's personal stacks) *which also have online articles* (i.e., their entire issues are available in both formats). Examples are: the prestigious journals Science, Nature, Cell and the Proceedings of the National Academy of Sciences of the U.S.A. (better known as PNAS).

Online-only journals have established a different way to cite. Since there are no paper pages, it suffices to have a number (e.g., an initial page) to indicate where the article starts. In addition, some journals offer downloadable citation files so writers can cite articles in a more standardized manner. For example, BMC Bioinformatics is only available online.

Zoppoli P, Morganella S, Ceccarelli M (2010) TimeDelay-ARACNE: Reverse engineering of gene networks from time-course data by an information theoretic approach. BMC Bioinformatics **11**:154.

Liang X-J, Xia Z, Zhang L-W, Wu F-X (2012) Inference of gene regulatory subnetworks from time course gene expression data. BMC Bioinformatics **13**(Suppl 9):S3.

Another way to cite online articles includes the Digital Object Identifier (DOI), which assigns a unique alphanumeric code to electronic papers. For example, here is the citation for an article from the Public Library of Science: Neglected Tropical Diseases:

Teixeira DE, Benchimol M, Crepaldi PH, de Souza W (2012) Interactive multimedia to teach the life cycle of *Trypanosoma cruzi*, the causative agent of Chagas disease. PLoS Negl Trop Dis **6**(8): e1749. doi:10.1371/journal.pntd.0001749

Of course, DOIs can be used to cite papers that are published both in print and online:

Brownlie, D. (2007). Toward effective poster presentations: An annotated bibliography. Eur J Marketing **41**:1245-1283. doi:10.1108/03090560710821161

The DOI offers an excellent alternative to everchanging URLs and they can be found as URLs themselves (e.g. www.mdpi.com/1422-0067/26/6/2536 will allow you to download the article below:

Valero MS, Gómez-Rincon C, Lopez V, Les F (2025) *Jasonia glutinosa* (L.) DC: Back in our pantries? A review of its pharmacological activity and mechanisms of action. Int J Mol Sci **26**:2536

Internet article (not Internet *journal* article)

Given the wide variety of web pages, one may follow a very simple style (e.g. that of the American Psychological Association, which is: Contributors' names (last edited date). *Title of resource*. Retrieved (date of retrieval) from web address for resource).

Angeli R, Wagner J, Lawrick E, Moore K, Anderson M, Soderland L, Brizee A (2010, May 5). *General format*. Retrieved from owl.english.purdue.edu/owl/resource/560/01/

The Edinborough Cell Wall Group. *Professor Stephen Fry's Research Interests*. Retrieved January 27, 2007, from the Word Wide Web: homepages.ed.ac.uk/sfry/research.html

Wolf A, Beegle D (1995) *Recommended soil tests for macronutrients: Phosphorus, potassium, calcium and magnesium*. In: Recommended soil testing Procedures for the Northeastern United States, 2nd Edition, Chapter 5, Delaware Cooperative Extension, Publications from the Soil Testing Laboratory. Northeastern Regional Publication N¡ 493.

Retrieved August 24, 2009, from the World Wide Web:

ag.udel.edu/extension/agnr/pdf/soiltesting/CHAP5-95.pdf.

A final note about using Internet citations: Many of the web pages you will find today will not be indefinitely available. Web page turnover is very fast! You should not cite more than 2-out-of-10 web pages per bibliography (or as indicated by your instructor) and that you save the corresponding HTML files for future reference (or print them as PDFs).
To cite online sources, you may check websites such as *The Write Source* of the Modern Language Association for examples of how to cite Internet references: www.thewritesource.com/mla/; the Purdue Online Writing Lab (https://owl.purdue.edu/); the Chicago Style Manual (https://www.chicagomanualofstyle.org/home.html), etc.

(F) Pipetting

Proficient pipetting is probably the most crucial part of this class, for three reasons:

- Accurate pipetting of very small volumes is crucial to the success of molecular projects. Mistakes during pipetting may cause your experiments to fail or to be irreproducible, and thus cause long delays and considerable expense.
- Pipettes are delicate pieces of equipment with high accuracy, which can easily be knocked off their calibration. Furthermore, they are expensive and maintenance expenses must be added to their cost.
- Pipettes must be kept clean; for example, dirty pipettes are the main source of contamination in PCR and other techniques.

You may have some experience with pipettes, but a refresher on pipetting is probably quite useful. In this section, we review the widely used Rainin Pipetman, models PR20, PR200, and PR1000, or similar.

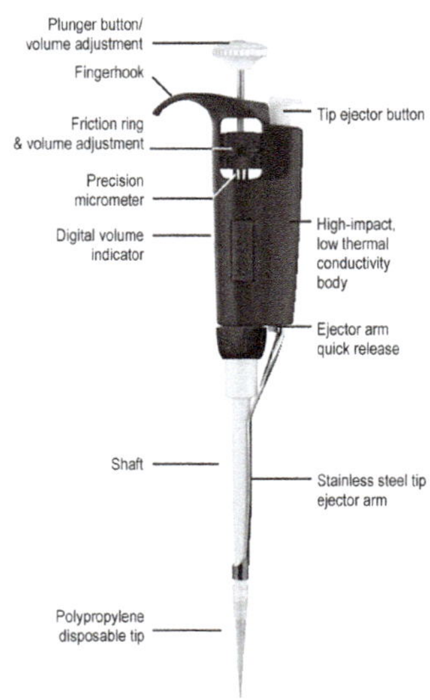

Figure 1. The Rainin Pipetman.

Volume Indicator

The volume indicator is read from top to bottom. Up to PR200, black digits indicate microliters and red digits tenths and hundredths of microliters. For PR1000 red digits indicate milliliters and black digits microliters (Table 1).

Table 1

PR20	PR200	PR1000
1	1	0
2	2	7
5	5	5
12.5 µL	125 µL	750 µL

Sample values, volume ranges, and smallest increments for each Rainin Classic model are given in Table 2.

Table 2

Model	Capacity (µL)	Recommended range (µL)	Smallest increment in µL
PR20	0 to 20	2 to 20	0.02
PR200	0 to 200	20 to 200	0.2
PR1000	0 to 1000	100 to 1000	2.0

Operation

1. Turn the plunger button or the volume adjustment knob until the volume indicator is 1/3 revolution above the desired setting, then turn slowly clockwise until the desired volume shows on the indicator.
2. *Always* dial down to the desired volume. This prevents mechanical backlash from affecting accuracy. If you pass the desired setting, turn the dial 1/3 revolution higher than desired and dial down to reset the volume. The friction ring prevents unintentional volume changes.

   Important note: The Rainin PR1000 pipettor has only 3 digits in the volume indicator window. A reading of "100" in the window is the maximum setting for the PR1000 equaling 1000 µL! **Do not attempt to dial the pipettor higher or you will break it!**
3. Attach a new disposable tip to the pipette shaft. Press into the tip with only enough force to make a positive airtight seal.
4. Press the plunger to the **first stop**. This part of the stroke is the volume displayed on the indicator.
5. Holding the pipette vertically, immerse the tip into the sample to the proper depth; see Table 3.
6. Allow the push button to return slowly to the **up** position. Never let it snap up! See Figure 2A below.
7. Pause briefly to ensure that the full volume of sample is drawn into the tip.
8. Withdraw the tip from the sample liquid. If any liquid remains on the outside of the tip, wipe it carefully with a lint-free tissue, avoiding the tip orifice.
9. To dispense sample, touch the tip end against the side wall of the receiving vessel and depress the plunger slowly to the first stop (see Figure 2B). Wait 1 s for PR2, PR10, PR20, PR100, PR200; wait 1–2 s for PR1000; wait 2–3 s for PR5000, PR10ML. Wait longer for viscous solutions.
10. Press the plunger to the **second stop** (bottom of stroke) to expel any residual liquid in the tip.
11. With the plunger fully pressed, withdraw the pipette from the vessel carefully, with the tip against the vessel wall.
12. Allow the plunger to return to the up position.
13. Discard the tip by depressing the tip ejector button. A fresh tip should be used for each sample to prevent sample carryover.

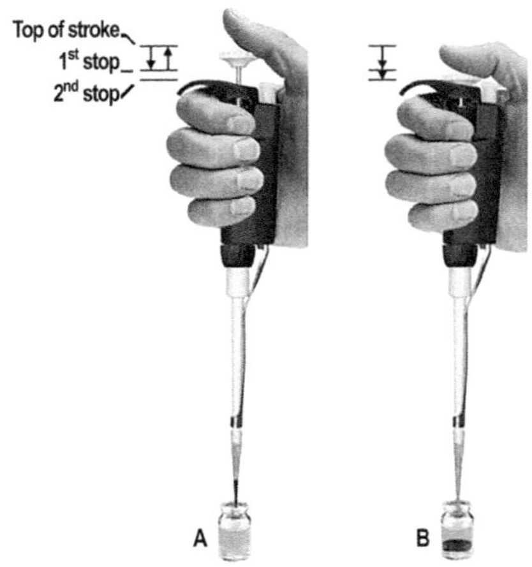

Figure 2. Operating Rainin Classic.

Tip Immersion Depth

The recommended depth for tip insertion into the sample for each Rainin Classic model is shown in Table 3.

Table 3

Model	Immersion depth
2 to 10 µL	1 to 2 mm
20 to 100 µL	2 to 3 mm
200 to 1000 µL	3 to 6 mm
5000 µL, 10 mL	6 to 10 mm

Tip immersion depth is important. If exceeded, the volume measured will be inaccurate, possibly out of specification.

Laboratory N° 1
Bioinformatics: Self-guided Internet-based Exercise on Databases for Storage and Data Mining

Purpose

This exercise aims to introduce you to some of the relevant databases and bioinformatics tools for examining and comparing different pieces of biological information stored in databases. Biological databases are an important resource (Maloney et al., 2010) for the study of biochemistry, molecular genetics, transmission genetics, cell biology, evolution and many other branches of the biological sciences.

Introduction

Biological databases contain enormous amounts of information about the sequences and structures of nucleic acids (DNA and RNA) and proteins; gene structures and chromosomes; metabolic pathways and enzymes; signaling mechanisms, etc. Some of them include software tools that can be used to analyze such data. Often, the software can be used directly through a web browser (web apps). Freestanding applications must be downloaded and installed on your computer or a local network.
Analyses of biological macromolecules, especially DNA, RNA and proteins, are based on the fundamental principle of gene expression, also known as the Central Dogma of Molecular Genetics (Fig. 3).

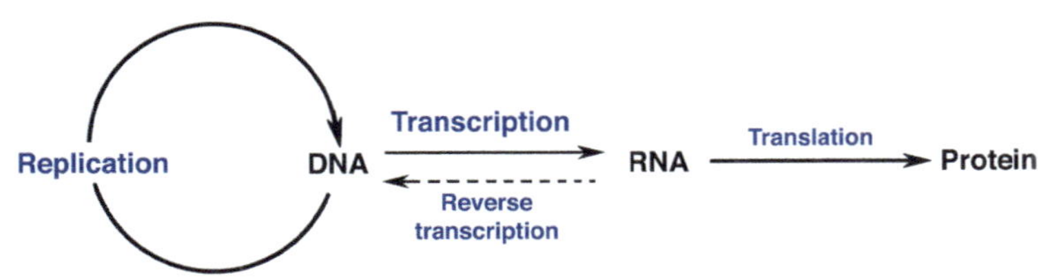

Figure 3. Central Dogma of Molecular Genetics.

The Internet hyperlinks are accessible via QR codes.

1 Finding Databases on the World Wide Web

Let's start by finding databases (Honts, 2003). You may use your cell phone and point to the QR codes in this document, and your assignment is to describe, in a short sentence, what the purpose of each website is. A home page usually has a brief description of what the purpose of the website is. Some titles are obvious (e.g. OMIM = Online Mendelian Inheritance in Man); others are not. For example, if you read the top of BLAST's first page, you'll find: "BLAST finds regions of similarity between biological sequences. The program compares nucleotide or protein sequences to sequence databases and calculates the statistical significance." BLAST stands for Basic Local Alignment Search Tool, and any biology student should become familiar with it. Click "Learn more" to find an expanded description.

DOI: 10.1201/9788743813224-1

1.1 General Databases and Tools for Bioinformatics Studies

National Center for Biotechnology Information
www.ncbi.nlm.nih.gov/
Brief description: [In your lab notebook or separate document file, write a summary of this database and its main usage.]

BLAST
blast.ncbi.nlm.nih.gov/Blast.cgi
Brief description: [In your lab notebook or separate document file, write a summary of this database and its main usage.]

PubMed
www.ncbi.nlm.nih.gov/pubmed
Brief description: [In your lab notebook or separate document file, write a summary of this database and its main usage.]

Online Mendelian Inheritance in Man (OMIM)
www.ncbi.nlm.nih.gov/omim
Brief description: [In your lab notebook or separate document file, write a summary of this database and its main usage.]

NCBI Conserved Domain Search
www.ncbi.nlm.nih.gov/Structure/cdd/wrpsb.cgi
Brief description: [In your lab notebook or separate document file, write a summary of this database and its main usage.]

CDART: Conserved Domain Architecture Retrieval Tool
www.ncbi.nlm.nih.gov/Structure/lexington/lexington.cgi?cmd=rps
Brief description: [In your lab notebook or separate document file, write a summary of this database and its main usage.]

European Bioinformatics Institute
www.ebi.ac.uk/
Brief description: [In your lab notebook or separate document file, write a summary of this database and its main usage.]

NOTES

Protein Data Bank
www.rcsb.org/
Brief description: [In your lab notebook or separate document file, write a summary of this database and its main usage.]

GenomeNet Database Resources
www.genome.jp/
Brief description: [In your lab notebook or separate document file, write a summary of this database and its main usage.]

1.2 Access Points for Integrated Suites of Sequence Analysis Tools

Multiple Protein Sequence Alignment
www.ncbi.nlm.nih.gov/tools/cobalt/cobalt.cgi?link_loc=BlastHomeLink
Brief description: [In your lab notebook or separate document file, write a summary of this database and its main usage.]

ExPASy
www.expasy.org/
Brief description: [In your lab notebook or separate document file, write a summary of this database and its main usage.]

Multiple protein or nucleic acid sequence alignments
www.ebi.ac.uk/Tools/msa/
Brief description: [In your lab notebook or separate document file, write a summary of this database and its main usage.]

PRABI (Pôle Rhône-Alpes Bioinformatics Center)
prabi.ibcp.fr/htm/site/web/app.php
Brief description: [In your lab notebook or separate document file, write a summary of this database and its main usage.]

1.3 Some Resources for Human Genomics

The Human Genome (NCBI)
www.ncbi.nlm.nih.gov/genome/guide/human/
Brief description: [In your lab notebook or separate document file, write a summary of this database and its main usage.]

Human Genome Browser Gateway (UCSC)
genome.ucsc.edu/cgi-bin/hgGateway
Brief description: [In your lab notebook or separate document file, write a summary of this database and its main usage.]

ENCODE
genome.ucsc.edu/ENCODE/
Brief description: [In your lab notebook or separate document file, write a summary of this database and its main usage.]

NOTES

1.4 Databases with Entire Genomic Sequences

National Center for Genome Resources
www.ncgr.org
Brief description: [In your lab notebook or separate document file, write a summary of this database and its main usage.]

J. Craig Venter Institute
www.jcvi.org/
Brief description: [In your lab notebook or separate document file, write a summary of this database and its main usage.]

Gramene: A Resource for Comparative Grass Genomics
www.gramene.org/
Brief Description: [In your lab notebook or separate document file, write a summary of this database and its main usage.]

Maize GDB (Maize Genetics and Genomics Database)
www.maizegdb.org/gbrowse
Brief Description: [In your lab notebook or separate document file, write a summary of this database and its main usage.]

1.5 Example of a Specialized Structure Prediction Tool

bio.tools COVID-19 Collection
bio.tools/t?domain=covid-19
Brief description: [In your lab notebook or separate document file, write a summary of this database and its main usage.]

1.6 Metabolic and Signaling Pathways

BioCyc (several organisms)
biocyc.org/
Brief description: [In your lab notebook or separate document file, write a summary of this database and its main usage.]

EcoCyc (*Escherichia coli*)
ecocyc.org/
Brief description: [In your lab notebook or separate document file, write a summary of this database and its main usage.]

Arabidopsis thaliana (thale cress)
www.arabidopsis.org/biocyc/
Brief description: [In your lab notebook or separate document file, write a summary of this database and its main usage.]

NOTES

Saccharomyces cerevisiae (brewer's yeast)
yeastgenome.org/
Brief description: [In your lab notebook or separate document file, write a summary of this database and its main usage.]

Danio rerio (zebra fish)
www.ensembl.org/index.html?redirect=no (scroll down to *Danio rerio*)
Brief description: [In your lab notebook or separate document file, write a summary of this database and its main usage.]

Mus musculus (mouse)
www.informatics.jax.org
Brief description: [In your lab notebook or separate document file, write a summary of this database and its main usage.]

Homo sapiens (human)
www.hmdb.ca
Brief description: [In your lab notebook or separate document file, write a summary of this database and its main usage.]

1.7 Additional Learning Resources

Kimball's Biology Textbook – One of the first online books ever
www.biology-pages.info
Brief description: [In your lab notebook or separate document file, write a summary of this database and its main usage.]

Taxonomy: In my opinion, Britannica is the best alternative to Wikipedia
www.britannica.com/science/taxonomyBrief description: [In your lab notebook or separate document file, write a summary of this database and its main usage.]

Phylogenetic trees: – Internet sources can be good, when properly curated
encyclopedia.thefreedictionary.com/phylogenetic tree
Brief description: [In your lab notebook or separate document file, write a summary of this database and its main usage.]

aleph0.clarku.edu/~djoyce/java/Phyltree/cover.html
Brief description: [In your lab notebook or separate document file, write a summary of this database and its main usage.]

bio.libretexts.org/Bookshelves/Introductory_and_General_Biology
Brief description: [In your lab notebook or separate document file, write a summary of this database and its main usage.]

NOTES

Google Scholar – Good engine search for scientific articles
scholar.google.com/schhp?hl=en&tab=ws
Brief description: [In your lab notebook or separate document file, write a summary of this database and its main usage.]

WolframAlpha – A math-oriented search engine, if you are in to numbers
www.wolframalpha.com/
Brief description: [In your lab notebook or separate document file, write a summary of this database and its main usage.]

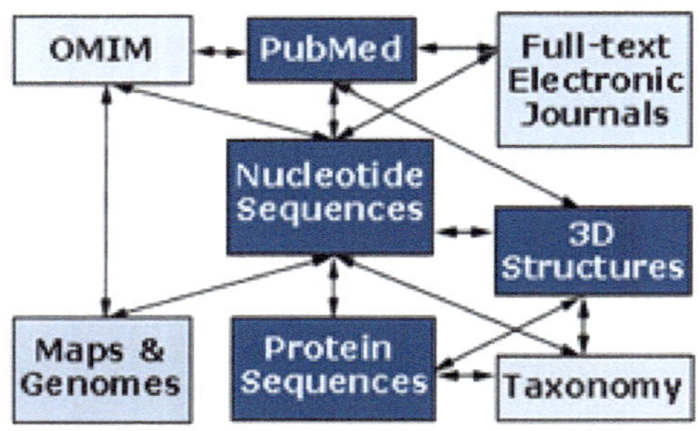

Figure 4. NCBI databases.

1.8 The National Center for Biotechnology Information

NCBI is a comprehensive network of databases that include information on nucleotidyl sequences (e.g. chromosomal DNA, mRNA, non-protein-coding RNAs), amino acyl sequences (proteins), taxonomy, genetically-based diseases (also known as "inborn errors of metabolism." Figure 4 illustrates the relationships among these different databases; you may want to continue exploring NCBI. The link www.ncbi.nlm.nih.gov/guide/all/#databases will take you to a comprehensive list of all NCBI databases (NCBI 2005).

2 Case Study: A Human Nucleotide Sequence

Specific Learning Objectives

- Describe what GenBank files are and be able to read them.
- Describe what FASTA format is and learn how to identify sequences in FASTA format.
- Become familiar with the BLAST program (check NCBI websites) and learn how to use it.

NOTE: Your instructor may decide to assign you a different sequence with respect to the one in this section. If this is the case, enter modifications to this document as necessary.

2.1

The nucleotide (short for "nucleotidyl residue") sequence on the following page comes from a human DNA sequencing project. You are given the task of identifying the location of this sequence within the human genome (Alaie et al., 2012). The problem is that the human genome is made up of 3 billion base pairs (bp). To check even 1000 bp by eye in search of this sequence is quite time-consuming (as you will find out shortly). Imagine if you had to check a billion nucleotides in a sequence!

Notice that the sequence provided on the next page, is in FASTA format, i.e., it does not include numbers, spaces or symbols. Notice that a name or designation for the sequence is written in the first line, preceded by the “>” symbol.

Start by scanning (by eye) the given sequence (4560 bp) in search of the location of the following short nucleotide stretches. Devise your own method.

(i) `CATTTCGTGTGGATCGCTCCTTTGCAAGTG` and (ii) `TTTAGAGAGAAGGCTGTCCTTAGTACCAGA`

Mark the locations on the FASTA format sequence found below (underline or use a highlighter) *or* on the electronic document, as requested by your instructor.

2.2

Please note the time at the beginning of your search and answer the following questions once you have located your sequence.

1. Describe the method you used to find the sequence stretches (visual comparison? computer-aided?).
2. How long did it take for you to find your sequence?

   Sequence (i) ___ (in *minutes*)

   Sequence (ii) ___ (in *minutes*)

2.3 BLAST

Let us explore the efficiency of using vast online databases and online search tools to locate and identify unknown nucleotide sequences. One such search tool is called BLAST (Basic Local Alignment Search Tool). This program compares a nucleotidyl (DNA, RNA) or amino acyl sequence (protein) of interest to online databases looking for regions of local similarity and calculates the statistical significance of matches. One such online database is NCBI's GenBank, which contains the sequences of at least three full-length human genomes and, being hosted by the National Library of Medicine (a branch of the National Institutes of Health), is free to the public.

Finding sequences of known (or putative) function in a database that have similarity to your sequence of interest may allow you to identify the gene family to which your sequence belongs or the functional significance of your sequence, if any. You will use a BLAST search to uncover information about an unknown sequence. Copy and paste the unknown sequence (either the one from last page or as provided by your section's instructor) onto a new Word document and save it in your computer's hard drive. Give it a title in the format Test_Sequence_LastName_FirstName.docx

- Go to NCBI BLAST website at blast.ncbi.nlm.nih.gov/Blast.cgi.
- In the resulting page, scroll down to Basic Blast and click on the link nucleotide blast. Copy the *first line* of the nucleotide sequence in the Word document and paste it in the “Enter Query Sequence” box. (The top line, preceded by the “>” sign, is a description of what the sequence is.)

```
>CELL_MOLEC TEST SEQUENCE (FASTA FORMAT)
GTAGTAGGTCTTTGGCATTAGGAGCTTGAGCCCAGACGGCCCTAGCAGGGACCCCAGCGCCCGAGAGACCATGCAGAGGT
CGCCTCTGGAAAAGGCCAGCGTTGTCTCCAAACTTTTTTTCAGCTGGACCAGACCAATTTTGAGGAAAGGATACAGACAG
CGCCTGGAATTGTCAGACATATACCAAATCCCTTCTGTTGATTCTGCTGACAATCTATCTGAAAAATTGGAAAGAGAATG
GGATAGAGAGCTGGCTTCAAAGAAAAATCCTAAACTCATTAATGCCCTTCGGCGATGTTTTTTCTGGAGATTTATGTTCT
ATGGAATCTTTTTATATTTAGGGGAAGTCACCAAAGCAGTACAGCCTCTCTTACTGGGAAGAATCATAGCTTCCTATGAC
CCGGATAACAAGGAGGAACGCTCTATCGCGATTTATCTAGGCATAGGCTTATGCCTTCTCTTTATTGTGAGGACACTGCT
CCTACACCCAGCCATTTTTGGCCTTCATCACATTGGAATGCAGATGAGAATAGCTATGTTTAGTTTGATTTATAAGAAGA
CTTTAAAGCTGTCAAGCCGTGTTCTAGATAAAATAAGTATTGGACAACTTGTTAGTCTCCTTTCCAACAACCTGAACAAA
TTTGATGAAGGACTTGCATTGGCACATTTCGTGTGGATCGCTCCTTTGCAAGTGGCACTCCTCATGGGGCTAATCTGGGA
GTTGTTACAGGCGTCTGCCTTCTGTGGACTTGGTTTCCTGATAGTCCTTGCCCTTTTTCAGGCTGGGCTAGGGAGAATGA
TGATGAAGTACAGAGATCAGAGAGCTGGGAAGATCAGTGAAAGACTTGTGATTACCTCAGAAATGATTGAAAATATCCAA
TCTGTTAAGGCATACTGCTGGGAAGAAGCAATGGAAAAAATGATTGAAAACTTAAGACAAACAGAACTGAAACTGACTCG
GAAGGCAGCCTATGTGAGATACTTCAATAGCTCAGCCTTCTTCTTCTCAGGGTTCTTTGTGGTGTTTTTATCTGTGCTTC
CCTATGCACTAATCAAAGGAATCATCCTCCGGAAAATATTCACCACCATCTCATTCTGCATTGTTCTGCGCATGGCGGTC
ACTCGGCAATTTCCCTGGGCTGTACAAACATGGTATGACTCTCTTGGAGCAATAAACAAAATACAGGATTTCTTACAAAA
GCAAGAATATAAGACATTGGAATATAACTTAACGACTACAGAAGTAGTGATGGAGAATGTAACAGCCTTCTGGGAGGAGG
GATTTGGGGAATTATTTGAGAAAGCAAAACAAAACAATAACAATAGAAAAACTTCTAATGGTGATGACAGCCTCTTCTTC
AGTAATTTCTCACTTCTTGGTACTCCTGTCCTGAAAGATATTAATTTCAAGATAGAAAGAGGACAGTTGTTGGCGGTTGC
TGGATCCACTGGAGCAGGCAAGACTTCACTTCTAATGGTGATTATGGGAGAACTGGAGCCTTCAGAGGGTAAAATTAAGC
ACAGTGGAAGAATTTCATTCTGTTCTCAGTTTTCCTGGATTATGCCTGGCACCATTAAAGAAAATATCATCTTTGGTGTT
TCCTATGATGAATATAGATACAGAAGCGTCATCAAAGCATGCCAACTAGAAGAGGACATCTCCAAGTTTGCAGAGAAAGA
CAATATAGTTCTTGGAGAAGGTGGAATCACACTGAGTGGAGGTCAACGAGCAAGAATTTCTTTAGCAAGAGCAGTATACA
AAGATGCTGATTTGTATTTATTAGACTCTCCTTTTGGATACCTAGATGTTTTAACAGAAAAAGAAATATTTGAAAGCTGT
GTCTGTAAACTGATGGCTAACAAAACTAGGATTTTGGTCACTTCTAAAATGGAACATTTAAAGAAAGCTGACAAAATATT
AATTTTGCATGAAGGTAGCAGCTATTTTTATGGGACATTTTCAGAACTCCAAAATCTACAGCCAGACTTTAGCTCAAAAC
TCATGGGATGTGATTCTTTCGACCAATTTAGTGCAGAAAGAAGAAATTCAATCCTAACTGAGACCTTACACCGTTTCTCA
TTAGAAGGAGATGCTCCTGTCTCCTGGACAGAAACAAAAAAACAATCTTTTAAACAGACTGGAGAGTTTGGGGAAAAAAG
GAAGAATTCTATTCTCAATCCAATCAACTCTATACGAAAATTTTCCATTGTGCAAAAGACTCCCTTACAAATGAATGGCA
TCGAAGAGGATTCTGATGAGCCTTTAGAGAGAAGGCTGTCCTTAGTACCAGATTCTGAGCAGGGAGAGGCGATACTGCCT
CGCATCAGCGTGATCAGCACTGGCCCCACGCTTCAGGCACGAAGGAGGCAGTCTGTCCTGAACCTGATGACACACTCAGT
TAACCAAGGTCAGAACATTCACCGAAAGACAACAGCATCCACACGAAAAGTGTCACTGGCCCCTCAGGCAAACTTGACTG
AACTGGATATATATTCAAGAAGGTTATCTCAAGAAACTGGCTTGGAAATAAGTGAAGAAATTAACGAAGAAGACTTAAAG
GAGTGCTTTTTTGATGATATGGAGAGCATACCAGCAGTGACTACATGGAACACATACCTTCGATATATTACTGTCCACAA
GAGCTTAATTTTTGTGCTAATTTGGTGCTTAGTAATTTTTCTGGCAGAGGTGGCTGCTTCTTTGGTTGTGCTGTGGCTCC
TTGGAAACACTCCTCTTCAAGACAAAGGGAATAGTACTCATAGTAGAAATAACAGCTATGCAGTGATTATCACCAGCACC
AGTTCGTATTATGTGTTTTACATTTACGTGGGAGTAGCCGACACTTTGCTTGCTATGGGATTCTTCAGAGGTCTACCACT
GGTGCATACTCTAATCACAGTGTCGAAAATTTTACACCACAAAATGTTACATTCTGTTCTTCAAGCACCTATGTCAACCC
TCAACACGTTGAAAGCAGGTGGGATTCTTAATAGATTCTCCAAAGATATAGCAATTTTGGATGACCTTCTGCCTCTTACC
ATATTTGACTTCATCCAGTTGTTATTAATTGTGATTGGAGCTATAGCAGTTGTCGCAGTTTTACAACCCTACATCTTTGT
TGCAACAGTGCCAGTGATAGTGGCTTTTATTATGTTGAGAGCATATTTCCTCCAAACCTCACAGCAACTCAAACAACTGG
AATCTGAAGGCAGGAGTCCAATTTTCACTCATCTTGTTACAAGCTTAAAAGGACTATGGACACTTCGTGCCTTCGGACGG
CAGCCTTACTTTGAAACTCTGTTCCACAAAGCTCTGAATTTACATACTGCCAACTGGTTCTTGTACCTGTCAACACTGCG
CTGGTTCCAAATGAGAATAGAAATGATTTTTGTCATCTTCTTCATTGCTGTTACCTTCATTTCCATTTTAACAACAGGAG
AAGGAGAAGGAAGAGTTGGTATTATCCTGACTTTAGCCATGAATATCATGAGTACATTGCAGTGGGCTGTAAACTCCAGC
ATAGATGTGGATAGCTTGATGCGATCTGTGAGCCGAGTCTTTAAGTTCATTGACATGCCAACAGAAGGTAAACCTACCAA
GTCAACCAAACCATACAAGAATGGCCAACTCTCGAAAGTTATGATTATTGAGAATTCACACGTGAAGAAAGATGACATCT
GGCCCTCAGGGGGCCAAATGACTGTCAAAGATCTCACAGCAAAATACACAGAAGGTGGAAATGCCATATTAGAGAACATT
TCCTTCTCAATAAGTCCTGGCCAGAGGGTGGGCCTCTTGGGAAGAACTGGATCAGGGAAGAGTACTTTGTTATCAGCTTT
TTTGAGACTACTGAACACTGAAGGAGAAATCCAGATCGATGGTGTGTCTTGGGATTCAATAACTTTGCAACAGTGGAGGA
AAGCCTTTGGAGTGATACCACAGAAAGTATTTATTTTTTCTGGAACATTTAGAAAAAACTTGGATCCCTATGAACAGTGG
AGTGATCAAGAAATATGGAAAGTTGCAGATGAGGTTGGGCTCAGATCTGTGATAGAACAGTTTCCTGGGAAGCTTGACTT
TGTCCTTGTGGATGGGGGCTGTGTCCTAAGCCATGGCCACAAGCAGTTGATGTGCTTGGCTAGATCTGTTCTCAGTAAGG
CGAAGATCTTGCTGCTTGATGAACCCAGTGCTCATTTGGATCCAGTAACATACCAAATAATTAGAAGAACTCTAAAACAA
GCATTTGCTGATTGCACAGTAATTCTCTGTGAACACAGGATAGAAGCAATGCTGGAATGCCAACAATTTTTGGTCATAGA
AGAGAACAAAGTGCGGCAGTACGATTCCATCCAGAAACTGCTGAACGAGAGGAGCCTCTTCCGGCAAGCCATCAGCCCCT
CCGACAGGGTGAAGCTCTTTCCCCACCGGAACTCAAGCAAGTGCAAGTCTAAGCCCCAGATTGCTGCTCTGAAAGAGGAG
```

- Leave the settings as they are, but on the **Choose Search Set** options, check the button **Genomic + transcript databases**. This will bring up the **Human genomic + transcript (Human G+T)** to the Database window. Scroll to the bottom of the page and click the **BLAST** button in the left-hand corner. Wait for results. Did your sequence find any matches in the human genome database?

  What could be the reason for this result?
- Now try a longer sequence. Copy the *top three lines* and paste this sequence into the "Enter Query Sequence" box and click **BLAST** again. Did your query match any sequence in the human genome database?

  If so, what match did it locate?
- Click on the first of the matches that your search yielded. This match should be with a sequence within GenBank. What is the name of this gene?
- What is the **Sequence ID**?
- Click on the Sequence ID link. This will take you to the entry for the DNA sequence and the predicted protein. What chromosome is this allele located in?

3 Conclusion

A fully processed messenger RNA (mRNA) contains nucleotide triplets in a particular sequence that are read from an *initiation codon* (AUG) up to one or two termination codons (out of three: UAG, UAA, UGA). The expression of a eukaryotic gene is controlled by DNA sequences called regulatory regions. The regulatory regions include the gene's promoter, which binds RNA polymerase once the transcription factors have bound the DNA and made that site accessible, and one or more enhancers that also bind transcription factors and contribute to the control of gene expression.

Usually, the expression of a gene can be modified if one of its regulatory regions undergoes a mutation. This mutation may be of immense significance, even if the change involves a single base substitution, since a transcription factor's recognition of the site is sequence specific. Mutations may involve more substantial changes to the gene's regulatory regions, such as multiple nucleotide deletions, or, as in the case of the gene under study in this lab, multiple nucleotide additions which may eventually result in the silencing of this gene.

The gene you searched encodes the *Homo sapiens* CF transmembrane conductance regulator (CFTR). The gene is a member of the ATP-binding cassette (ABC) transporter superfamily. The encoded protein functions as a chloride channel, making it unique among members of this protein family, and controls ion and water secretion and absorption in epithelial tissues. Channel activation is mediated by cycles of phosphorylation of the regulatory domain, ATP-binding by the nucleotide-binding domains, and ATP hydrolysis.

Mutations in this gene cause cystic fibrosis, the most common lethal genetic disorder in populations of Northern European descent. The most frequently occurring mutation in cystic fibrosis, ΔF508, results in impaired folding and trafficking of the encoded protein (Morral et al., 1994). Multiple pseudogenes have been identified in the human genome (Tsui et al., 1992).

4 Discussion

You are given a sequence of DNA and told that it is human. You are asked to find out its identity and whether it has similarity to sequences in other organisms. Please describe the bioinformatics tool, the database, and the procedure you would use to find such information. Give two possible outcomes of your search.

Questionnaire
Lab 1. Bioinformatics: Self-guided Internet-based Exercise on Databases for Storage and Data

- Consider the sequence you searched using the BLAST program. Assuming that this is the *wild type* (unmutated form), how would you find possible mutants using BLAST?

  Explain your reasoning.

  A database containing all known nucleotide sequences exists and is called "nucleotide collection (nr/nt)." This database can be found on the BLAST site under "Choose Search Set." At "Database" you will see that the "Human Genome + transcript" is selected. Select "Others" instead and you will find that the "nucleotide collection (nr/nt)" database is automatically selected. Run your search against this vast database.
- What result would you predict if we searched that sequence against all known sequences?
- How do your results differ from the original search?
- Describe the capabilities of a BLAST search.
- What could be the possible limitations of a BLAST search?
- BLAST is often nicknamed "the Google of DNA search tools." Compare a BLAST search to a Google search and list at least one possible similarity and at least one possible difference.

NOTES

NOTES

Laboratory N° 2
Biological Buffers I. Titration of the Amino Acid Glycine

Purpose

The exercises in this laboratory are designed to illustrate the ionization properties of weak acids and amino acids (Figure 5) as a basis for understanding the properties of proteins and their use as biological buffers. You will calculate the components and prepare a buffer of a known pH. This exercise is intended to provide you with practical experience in basic laboratory procedures and illustrate the buffering properties of common biochemical solutions.

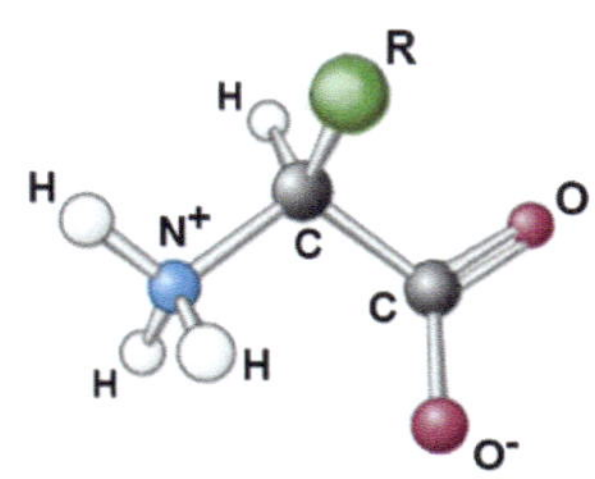

Figure 5. Generalized structure of an L-amino acid.

Importance of pH Regulation in Living Organisms

Most intracellular processes take place in aqueous environments. Water is an amphoteric substance that serves as a proton donor (acid) or a proton acceptor (base). It exists in equilibrium as represented in Equation (1):

$$H_2O \rightleftharpoons H^+ + OH^- \quad (1)$$

In *pure* water, $[H^+] = [OH^-] = 10^{-7}$ M. However, the concentration of H^+ and OH^- in aqueous solutions can vary greatly because many solutes contribute (acids) or absorb (bases) protons. Although the ratio $[H^+]/[OH^-]$ can vary greatly in aqueous solutions, the product of the proton and hydroxide ion concentration in aqueous solutions is always 10^{-14} M (i.e. $[H^+]$ X $[OH^-] = 10^{-7} \times 10^{-7} = 10^{-14}$ M). Therefore, $[OH^-]$ of an aqueous solution can be calculated if $[H^+]$ is determined experimentally and vice versa.
As a convenient way of expressing the concentration of $[H^+]$ in a solution, chemists use the term pH. The alkalinity or acidity of a solution is expressed in pH units. pH is defined as the negative logarithm of the concentration of protons in a solution:

$$pH = -\log[H^+] \quad (2)$$

$$\text{or } pH = \log\frac{1}{[H^+]} \quad (3)$$

Therefore, with $[H^+] = 10^{-7}$ M the pH of pure water should be 7. However, many acidic and basic solutes inside of the cell react with the protons or hydroxyl groups. This could potentially shift the equilibrium of the water equilibrium reaction above and therefore dramatically change the intracellular pH. As one might imagine, such wide fluctuations in cellular pH would seriously affect normal physiological processes.

DOI: 10.1201/9788743813224-2

In order to maintain a constant pH, cells utilize natural buffering agents that absorb or contribute protons to maintain a constant pH. In fact, some organelles have sub-compartments that lead to the formation of pH gradients, which form part of important physiological activities (e.g., ATP production across the inner membrane (cristae) of the mitochondrion or the thylakoids in the chloroplast). In addition, to study biological reactions in vitro, it is necessary for investigators to use artificial ("biological") buffers to maintain a constant pH during a reaction, cell manipulation, or when culturing tissues and cells.

Buffers

A *buffer* is a solution that resists pH change upon the addition of acid or base (Figure 6). All laboratory pH buffers can be thought of as weak acids and their dissociation in water can be described by the following equation:

$$\mathrm{HA} \leftrightarrows \mathrm{A}^- + \mathrm{H}^+ \tag{4}$$

where HA = any weak acid. The degree of the dissociation of HA into A^- and H^+ can be described by K_a, an equilibrium constant:

$$\mathrm{K_a} = \frac{[\mathrm{A}^-][\mathrm{H}^+]}{[\mathrm{HA}]} \tag{5}$$

By taking the reciprocal of both sides of Equation (5):

$$\frac{1}{\mathrm{K_a}} = \frac{[\mathrm{HA}]}{[\mathrm{A}^-][\mathrm{H}^+]} \tag{6}$$

and by taking the log of both sides:

$$\log\frac{1}{\mathrm{K_a}} = \log\frac{[\mathrm{HA}]}{[\mathrm{A}^-][\mathrm{H}^+]} \tag{7}$$

From the definition of pH in Equation (3), in a similar manner we may establish:

$$\log\frac{1}{\mathrm{K_a}} = \mathrm{pK_a} \tag{8}$$

Therefore:

$$\mathrm{pK_a} = \mathrm{pH} + \log\frac{[\mathrm{HA}]}{[\mathrm{A}^-]} \tag{9}$$

By reorganizing the terms in Equation (9), we can derive an expression that describes the pH of a solution containing a solute with a known pK_a:

$$\mathrm{pH} = \mathrm{pK_a} + \log\frac{[\mathrm{A}^-]}{[\mathrm{HA}]} \tag{10}$$

This is the *Henderson–Hasselbalch equation* and it will be useful throughout the course and during your scientific career to understand the importance of pH in biological systems and as a tool in the laboratory for making buffer solutions.

Buffering Capacity of Amino Acids

Amino acids are the basic building blocks of protein molecules. Amino acids present in proteins have a carboxyl group and an amino group bound to the same carbon atom (the α-carbon). Together with a hydrogen atom and a lateral (R) group, the α-COOH and α-NH_2 groups are present in the L-configuration (Figure 5).
In free amino acids, the α-COO^- and α-NH_3^+ groups are weak acid groups that ionize in

NOTES

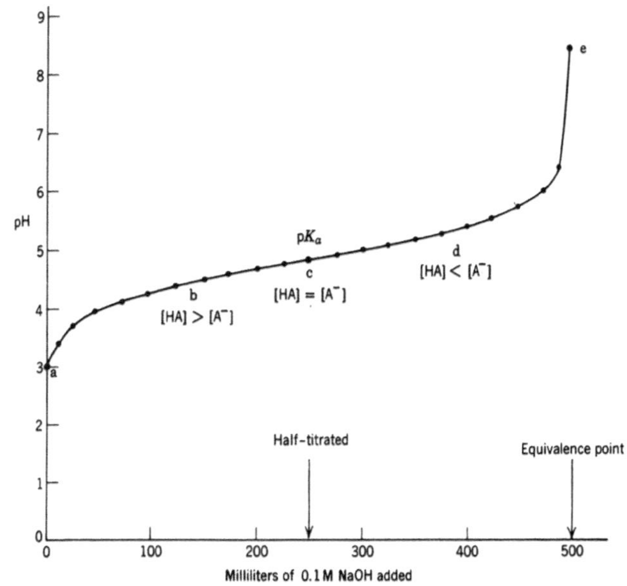

Figure 6. Titration of a weak monoprotic acid (such as acetic acid), at 0.1 M, with a strong base (i.e. 0.1 M NaOH); pK_a = 4.8. (Adapted from Siegel, 1976.).

aqueous solutions by donating a proton. These groups have pK_a values between 2 and 3 and 9 to 10, respectively, depending on the R group. In addition, some amino acids (e.g., aspartic acid and lysine) have lateral side chains that also act as weak acids and bases (see pK_a values of R groups in Figure 7). With the exception of the α-amino and α-carboxy terminal groups, the α-COO^- and α-NH_3^+ from the aminoacyl residues in proteins are engaged in peptide bonds; therefore, the ability of the ionizable lateral groups to donate and accept protons is essential to the chemical properties of proteins, and the ionization of these groups often take part in the formation of protein structure and the catalytic activities of enzymes (Siegel, 1976)

pH Measurement

The pH of a solution can be measured using diverse methods. The two most common methods are pH paper and the pH meter. The so-called pH paper is impregnated with a mixture of dyes. The dyes in the paper will change color depending on the pH of the surrounding solution. The pH of an unknown solution can be determined by placing a drop of a solution on the paper and comparing the paper's color to a set of color standards generated from known pH values. This method is accurate only within 0.5 pH units because of the human error involved when comparing the colors of the unknown vs. the standard papers. In consequence, pH paper cannot be used to *quantitatively* determine the pH of a solution. However, pH paper is useful in situations where the researcher needs to quickly determine the approximate pH of a solution or when a pH meter is not available.

The pH meter is used to determine the exact pH of a solution, and it consists of a voltmeter attached to an electrode that is permeable to H^+ ions. Submerging the electrode in a solution will result in a voltage reading that corresponds to the concentration of H^+ in the solution. The display of the voltmeter is usually altered to give a direct reading of the pH rather than a voltage. The pH meter will give an exact measurement of $[H^+]$ in a solution. In this exercise, you will use a pH meter to prepare solutions of a given pH. The use of the instruments and their calibration is described in a guide that should be placed next to each instrument.

Please be careful not to hit the electrode if you are using a magnetic stir bar.

The α-L-Amino Acids

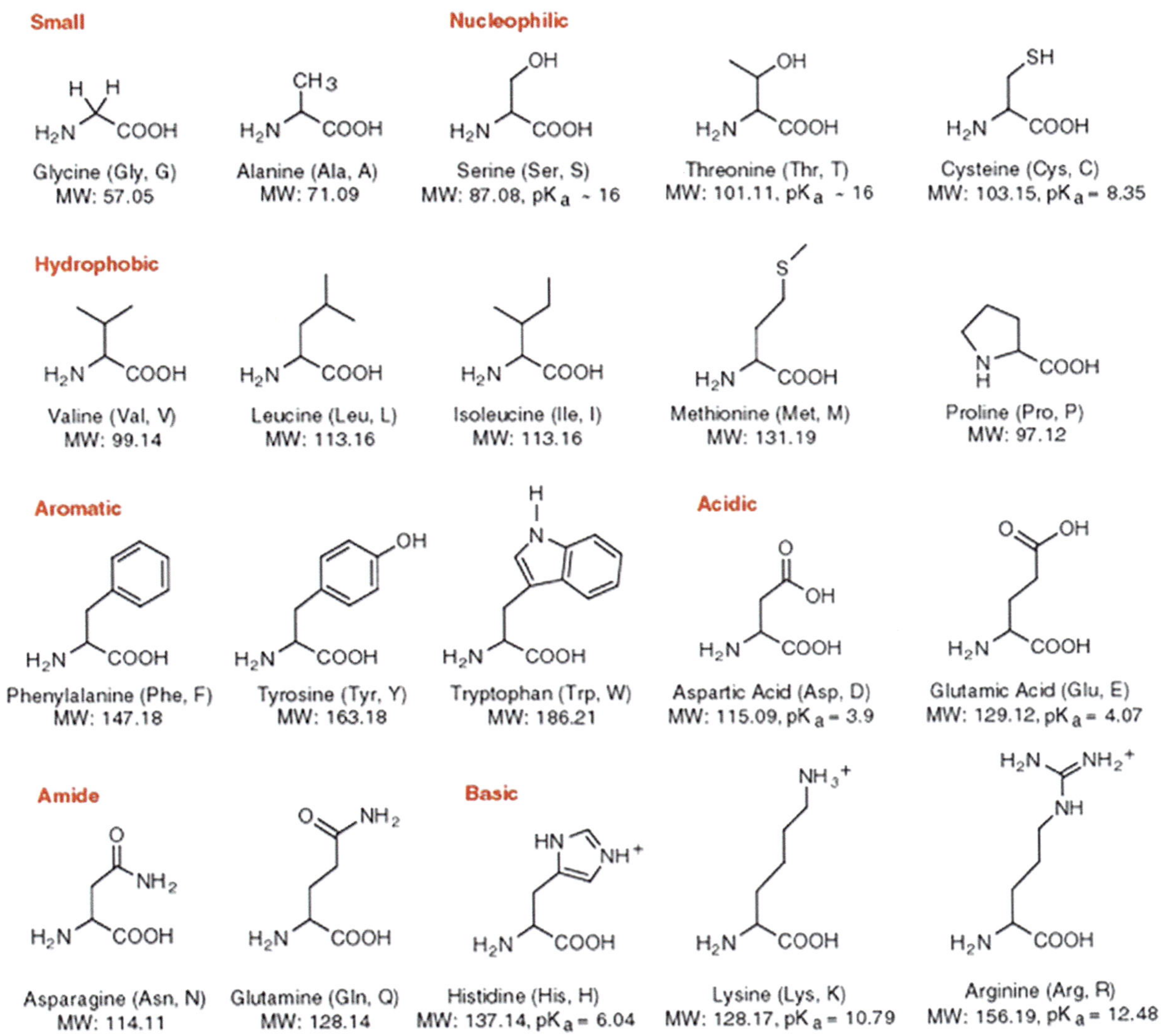

Figure 7. Proteins are composed of 19 amino acids and one imino acid (Pro). MW = molecular weight. The indicated pK_a corresponds to the R lateral group. Adapted from New England Bio Labs (2014).

Titration of the Amino Acid Glycine

To illustrate the ionization properties of the functions groups of amino acids, you will perform a titration of the α-COO^- and α-NH_3^+ groups of glycine (Gly), the simplest amino acid, which has a non-ionizable R group (H; see Figure 7). From your titration data, you will determine the pK_a values of these two functional groups.

1. Calibrate the pH meter according to the instructions supplied with the instrument and using the standard provided by the instructor (red, pH 4; yellow, pH 7; blue, pH 10).

2. Transfer 50 mL of a 0.1 M solution of Gly to a clean 250 mL beaker and add 50 mL of deionized water (diH_2O). Place a stir bar in the beaker and place the beaker on a stir plate next to the pH meter.
3. Rinse the electrode of the pH meter and place it in the Gly solution. Stir the solution very gently and read the initial pH.
4. Fill a 25 mL burette with a solution of 1 M NaOH and place it over the Gly solution.
5. Begin titrating the Gly solution by adding 0.2 mL increments of 1 M NaOH while gently stirring the solution. Record the precise volume of NaOH added at each step. Be sure that the electrode remains submerged during the procedure. Stop between the addition of each increment of 1 M NaOH and record the pH. Be sure that you allow the pH to stabilize before you record the value on your notebook.
6. Continue the titration until pH reaches 12 or 13.
7. Tabulate the data on your notebook. *Follow the format established in Table 4.* Calculate the equivalents of base added at each step of the titration and calculate the accumulated equivalents at each step.

Table 4

Increment (mL NaOH)	Vol. of 1 M NaOH (mL)	Equivalents added at increment	Accumulated equivalents	pH
1				
2				
3				
4				
…	…	…	…	…
n				

8. An equivalent refers to the moles of OH^- or H^+ added with each addition of NaOH. It is determined by multiplying the volume of NaOH added by the concentration of NaOH:
Equivalents $= 1$ mole/L $\times$ 0.0002 L $= 2 \times 10^{-4}$ (moles/L) $\cdot$ L $= 2 \times 10^{-4}$ moles.
9. Plot the accumulated equivalents of NaOH added vs. the pH on a linear graph similar to that shown in Figure 1 (part A). Label your graph properly, i.e., the independent variable is NaOH equivalents; the dependent variable is pH.
10. From your plot, estimate the pK_a values for the α-COOH and α-NH_3^+ groups of Gly.

NOTES

Questionnaire
Lab 2. Titration of the Amino Acid Glycine

1. *Following the format established in the table on the previous page*, tabulate the titration data on your notebook and calculate the equivalents of base added at each step of the titration
2. Calculate the accumulated equivalents at each step using the formula on your lab manual
   NaOH Equivalents = 1 mole/L $\times$ 0.0002 L = 2×10^{-4} (moles/L) $\cdot$ L = 2×10^{-4} moles.
3. Plot the pH (Y-axis or ordinate) vs. accumulated equivalents of NaOH added (X-axis or abscissa) on a linear graph similar to that shown in Figure 1 (part A). Label your graph properly, i.e., the independent variable is NaOH equivalents; the dependent variable is pH.
4. From your plot, estimate the pK_a values for the α-COOH and α-NH_3^+ groups of Gly. Use arrows to indicate these pK_a values on the curve.
5. Make a photocopy of the table and the plot and attach both of them to this report.
6. What is the ratio of HOOC-Gly-NH_3^+/$^-$OOC-Gly-NH_3^+ at pK_{a1}?
7. What is the ratio of $^-$OOC-Gly-NH_3^+/$^-$OOC-Gly-NH_2 at pK_{a2}?
8. We will use glycine buffers in experiments later in the semester. Why do different proportions of the ionic species of glycine at certain pH values allow this molecule to be used as pH buffer?
9. Would glycine be a better buffer in a higher animal circulatory system than phosphates or bicarbonate/carbonate?

❑ Yes ❑ No

10. Explain your answer to the question above.

NOTES

NOTES

Laboratory N° 3
Biological Buffers II. Preparation of a Phosphate Buffer

When the concentrations of the conjugate acid and base of a buffering compound are equal, then $pH = pK_a$. This is the point of maximum *buffering capacity* of the solution. This is most obvious when a titration curve of the substance is generated by the titration of a buffer of a known concentration with varying amounts of a strong acid or base. For a monoprotic buffer (i.e. made from a weak acid that contributes only a single proton, such as CH_3COOH) the titration curve will look like that in Figure 6 (in Lab 2), but the volume of base required will vary depending on the initial concentration of the acid. For acids whose physical state is liquid in laboratory conditions, one must factor in their density ($\rho = m/v$) and, in some instances, their purity (if it is considerably lower than 100% v/v).

Blood Buffers

An excellent example of buffer capacity is found in the blood plasma of mammals, which has a remarkably constant pH (Horton et al., 2002) . Consider the results of an experiment that compares the addition of an aliquot of a strong acid to a volume of plasma with a similar addition of strong acid to either physiological saline (0.15 M NaCl) or water. When 1 mL of 10 M HCl is added to 1 L of physiological saline or water that is initially at pH 7.0, the pH is lowered to 2.0 (in other words, H^+ from HCl is simply diluted to 10^{-2} M). However, when 1 mL of 10 M HCl is added to 1 L of human blood plasma at pH 7.4, the pH is lowered to only 7.2 (impressive evidence for the effectiveness of physiological buffering).
The bicarbonate/carbon dioxide (HCO_3^-/CO_2) system is one of the two major blood buffers, the other being the protein hemoglobin. Carbonic acid (H_2CO_3) ionizes as a typical weak diprotic acid (Siegel, 1976):

$$H_2CO_3 \rightleftharpoons HCO_3^- + H^+ \rightleftharpoons CO_3^{2-} + H^+$$

However, most of the conjugate acid[1] dissolved in blood and cytoplasm is present as CO_2, not H_2CO_3. The dissolved CO_2 is in equilibrium with CO_2 in the gas phase:

$$CO_2\ (gas) \overset{k}{\rightleftharpoons} CO_2\ (dissolved) + H_2O \overset{k_{eq1}}{\rightleftharpoons} H_2CO_3 \overset{k_{a1}}{\rightleftharpoons} HCO_3^- + H^+ \overset{k_{a2}}{\rightleftharpoons} CO_3^{2-} + H^+$$

The equilibrium between CO_2 (gas) and CO_2 (dissolved) is given by:

$$\boxed{[CO_2]_{dissolved} = k\left(P_{CO_2}\right)}$$

That is, the concentration of dissolved CO_2 is directly proportional to the partial pressure of CO_2 (P_{CO_2}) in the gas phase. At 37 °C and an ionic strength of 0.5, $k = 3.01 \times 10^{-5}$ when P_{CO_2} is expressed in terms of mm Hg.
The equilibrium constant for the reaction $CO_2\ (dissolved) + H_2O \rightleftharpoons H_2CO_3$ is about 5×10^{-3}:

$$K_{eq1} = \frac{[H_2CO_3]}{[CO_2]_{dissolved}} 3 = 5 \times 10^{-3}$$

[1]Protons become part of hydronium ions (H_3O^+) every time each one of these chemical species dissociates.

DOI: 10.1201/9788743813224-3

Thus, the overall equilibrium constant between dissolved CO_2 and $HCO_3^- + H^+$ is given by:

$$K_a' = \frac{[HCO_3^-][H^+]}{[CO_2]_{dissolved}} = K_{eq1} \times K_{a1} = (5 \times 10^{-3})(1.58 \times 10^{-4}) = 7.9 \times 10^{-7}$$

$$\therefore pK_a' = 6.1$$

At any pH:

$$pH = 6.1 + \log\frac{[HCO_3^-]}{[CO_2]} \text{ and } pH = 6.1 + \log\frac{[HCO_3^-]}{(3.01 \times 10^{-5})\,P_{CO_2}}$$

For all practical purposes, a bicarbonate buffer can be considered to be composed of HCO_3^- (conjugate base) and dissolved CO_2 (conjugate acid).

The pH of blood is maintained at about 7.4. If the pK_a of CO_2 is 6.1, how can the HCO_3^-/CO_2 system help buffer blood at pH 7.4? (Remember that a buffer is supposed to be effective only in the region of its pK_a.) In vivo, the HCO_3^-/CO_2 is an open system in which $[CO_2]_{dissolved}$ is maintained constant. Any excess CO_2 produced by the reaction $HCO_3^- + H^+ \rightleftharpoons H_2O + CO_2$ is expelled by the lungs. In contrast, the usual laboratory-buffer is a closed system: The concentration of conjugate acid increases when H^+ reacts with the conjugate base.

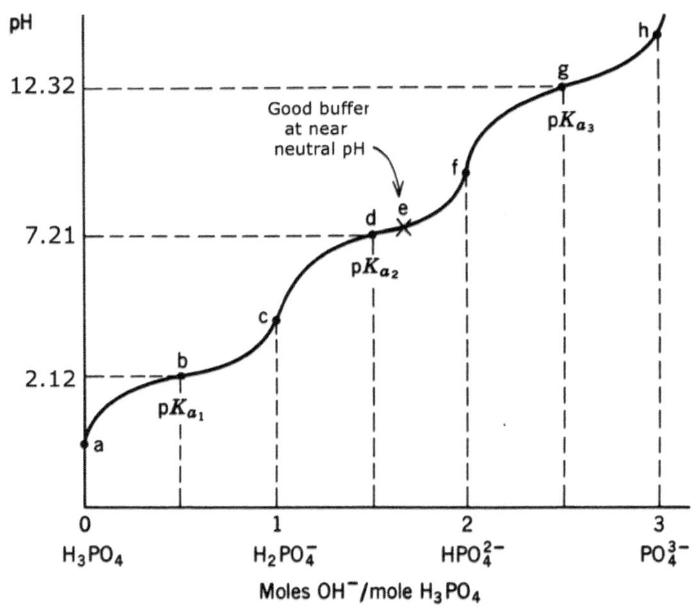

Figure 8. Titration of phosphoric acid, a multiprotic molecule, with a strong base. The initial point, where 100% of the H_3PO_4 molecules are not ionized, is indicated by **a**. Equivalence points for each ionizable group are indicated by **c**, **f**. and **h**. (Adapted from Siegel, 1976).

NOTES

Multiprotic Buffers

Many laboratory buffers are multiprotic (e.g. phosphoric acid, glycine). Therefore, these buffers have multiple pK_a values (cf. Figure 8), corresponding to the dissociation of protons from each functional group. For phosphoric acid the proton dissociation equations can be written as:

$$\begin{aligned}
H_3PO_4 &\rightleftharpoons H_2PO_4^- + H^+ \quad pK_{a1} = 2.12\\
H_2PO_4^- &\rightleftharpoons HPO_4^{2-} + H^+ \quad pK_{a2} = 7.21\\
HPO_4^{2-} &\rightleftharpoons PO_4^{3-} + H^+ \quad pK_{a3} = 12.32
\end{aligned}$$

To use the Henderson–Hasselbalch equation with multiprotic buffers, one must select the pK_a and the corresponding dissociation equation that is closest to the desired pH of the solution.

Examples

1. Acetic acid (CH_3COOH) has a pK_a of 4.8. How many mL of 0.1 M acetic acid and 0.1 M sodium acetate ($CH_3COO^-Na^+$)are required to prepare 1 liter of 0.1 M acetate buffer, pH 5.8 (Horton et al., 2002, pp. 38-42)?

   Substitute the values for pK_a and the desired pH into Equation (10):

$$5.8 = 4.8 + \log \frac{[CH_3COO^-]}{[CH_3COOH]}$$

   Solve for the ratio of acetate to acetic acid:

$$\begin{aligned}
&\log \frac{[CH_3COO^-]}{[CH_3COOH]} = \frac{5.8 - 4.8}{1.0} = \\
&\frac{[CH_3COO^-]}{[CH_3COOH^-]} = \text{antilog } 1 = 10\\
&\therefore [CH_3COO^-] = 10\,[CH_3COOH]
\end{aligned}$$

   For each volume of acetic acid, 10 volumes of acetate must be added (making a total of 11 volumes of the two ionic species). Multiply the proportion of each component by the desired volume (in this case, 1000 mL) and mix:

$$\begin{aligned}
0.1\text{ M } CH_3COOH \text{ needed:} \quad & 1/11(1000\text{ mL}) = \quad 91\text{ mL}\\
0.1\text{ M } CH_3COO^-Na^+ \text{needed:} \quad & 10/11(1000\text{ mL}) = \frac{909\text{ mL}}{1000\text{ mL}}
\end{aligned}$$

   Note that when the ratio of $[A^-]$ to [HA] is 10:1, the pH is exactly one unit above the pK_a. If the ratio were 1:10, the pH would be one unit below the pK_a.

2. Calculate mL of glacial acetic acid (17.6 N; assume 99.7% purity)[2] and g of sodium acetate (f.w. = 82 g/mole) that is required to make 100 ml of 0.2 M buffer at pH 3.9.

   Again, substituting in Equation (10) with the desired pH:

$$3.9 = 4.8 + \log \frac{[CH_3COO^-]}{[CH_3COOH]}$$

[2]Important information for CH_3COOH: f.w. (formula weight) = 60.05 g/mole; ρ = 1.05 g/cm^3.

Let $[CH_3COOH] = x_1$ mole/L, then $[CH_3COO^-] = (0.2 - x_1])$ mole/L.

Therefore, pH = pKa + log $(0.2 - x_1/x_1)$ and 3.9 = 4.8 + log $(0.2 - x_1/x_1)$ ∴ −0.9 = log $(0.2 - x_1/x_1)$ and $x_1 = 0.178$ mole/L = $[CH_3COOH]$.

The concentration of acetate ion can be found from the equation:

moles of CH_3COOH/L + moles of CH_3COO^-/L= 0.2 moles/L
or $x_1 + x_2 = 0.2$ mole/L where x_2 represents moles/L of CH_3COO^-.

This expression indicates that [acid] and [conjugated base] must account for the final concentration of the buffering chemical species in the solution.

Because concentrations are not additive, the number of moles for acid and conjugated base are added instead, but relative to a volume (i.e. per liter). Also, recall that 1 mole/L = f.w. (g)/L and, for monoprotic acids, 1 M = 1 N.

From the equation above:

0.178 mole/L + x_2 = 0.2 mole/L

∴x_2 = 0.2 mole/L – 0.178 mole/L → x_2 = 0.022 mole/L of $[CH_3COO^-]$

For 100 mL of solution:

0.1 L × 0.178 mole/L (L/17.6 mole) = 0.00101 L = **1.01 mL of CH_3COOH**

and 0.1 L × 0.022 mole/L × 82 g/mole = **0.18 g CH_3COONa**

Mix the above in ~25 mL of water and bring the volume to 100 mL in a volumetric flask.

3. Calculate the pH of a solution that was prepared by dissolving 1.83 g of KH_2PO_4 (anhydrous; f.w. = 136) and 1.16 g of K_2HPO_4 (anhydrous; f.w. = 174) in 0.2 L of water.

   Solution: The applicable dissociation equation for this solution corresponds to $pK_{a2} = 7.21$ (see Figure 4). Therefore:

$$pH = pK_{a2} + \log \frac{[K_2HPO_4]}{[KH_2PO_4]}$$

   $[K_2HPO_4]$ = 1.16 g × mole/174 g × 1/0.2 L = 0.033 M

   [KH2PO4] = 1.83 g × mole/136 g × 1/0.2 L = 0.067 M

   pH = 7.21 + log (0.033/0.067) ∴ pH = 6.92

4. The pH of a sample of arterial blood is 7.42. Upon acidification of 10 mL of the blood, 5.91 mL of CO_2, corrected for standard temperature and pressure (S.T.P.), is produced. Calculate (a) the total concentration of dissolved CO_2 in the blood $[CO_2 + HCO_3^-]$, (b) the concentration of dissolved CO_2 and HCO_3^-, and (c) the partial pressure of the dissolved CO_2 (in mm Hg).

   (a) First, calculate the number of moles of CO_2 represented by 5.91 mL at S.T.P. One mole of a "perfect gas" occupies 22.4 L at S.T.P.; for CO_2, the experimental value is 22.26 L.

$$\therefore \frac{5.91 \times 10^{-3}\ \text{L}}{22.6\ \text{L/mole}} = 26.5 \times 10^{-5} \text{ moles of } CO_2$$

   This amount of CO_2 came from 10 mL (0.01 L = 1×10^{-2} L) of blood

$$\therefore \text{concentration of "total } CO_2\text{"} = \frac{26.5 \times 10^{-5} \text{ moles}}{1 \times 10^{-2}\text{ L}} = 2.65 \times 10^{-2}\text{M}$$

(b)

$$\text{pH} = K_{a1} + \log\frac{[HCO_3^-]}{[CO_2]}$$

$$7.42 = 6.1 + \log\frac{[HCO_3^-]}{[CO_2]} \quad 1.32 = \log\frac{[HCO_3^-]}{[CO_2]}$$

$$\frac{[HCO_3^-]}{[CO_2]} = \text{antilog } 1.32 = 20.89\frac{20.89}{1}$$

$$[HCO_3^-] = \frac{20.89}{21.89} \times 2.65 \times 10^{-2}\text{ M} = 2.53 \times 10^{-2}\text{ M}$$

$$[CO_2] = \frac{1}{21.89} \times 2.65 \times 10^{-2}\text{ M} = 1.21 \times 10^{-3}\text{ M}$$

(c) $[CO_2]_{\text{dissolved}} = k\,(PCO_2)$ mm Hg

$$P_{CO_2} = \frac{[HCO_3^-]}{[CO_2]} = \frac{1.21 \times 10^{-3}}{3.01 \times 10^{-5}} = 40.22 \text{ mm Hg}$$

Experimental Procedure

1. Prepare 100 mL of 0.3 M phosphate buffer, pH 7.5. Start with solid Na_2HPO_4 and KH_2PO_4 and diH_2O. Use the Henderson–Hasselbalch equation to calculate the number of moles/100 mL of each phosphate salt that you need to prepare the solution. Convert the molar amount to g using the formula weight given on the chemical containers and weigh out the appropriate amount of each component on the balance.
2. Set up a 100 mL beaker on a stir plate. Add 80 mL of diH_2O[3] and a stir bar and start stirring the liquid. Dissolve the solid phosphate salts by slowly adding each powder to the beaker. Allow the solid to dissolve completely. When they are dissolved, adjust the volume to 100 mL using a 100 ml graduated cylinder (Erlenmeyer flasks and beakers have less accurate volumes than cylinders).
3. Transfer the solution back to the beaker and measure the pH of the solution using the pH meter. Does it agree with the desired pH of 7.5? If not, can you explain why?
4. One of the most common buffers used in experimental biology is called phosphate buffered saline (PBS). There are many formulations for PBS. For example: weigh 8 g sodium chloride (NaCl), 0.2 g potassium chloride (KCl), 2.17 g sodium phosphate dibasic (Na_2HPO_4)[4], 0.2 g potassium phosphate monobasic (KH_2PO_4); dissolve in 80 mL of diH_2O. Measure the pH; adjust the pH to 7.5 if necessary and add dH_2O to 100 mL. What are the final concentrations of each ion in your PBS?

$$[PO_4^{3-}] = \_\_\text{ mM} \qquad [Na^+] = \_\_\text{ mM} \qquad [K^+] = \_\_\text{ mM} \qquad [Cl^-] = \_\_\text{ mM}$$

[3]diH_2O = deionized water; dH_2O = distilled water, ddH_2O = double-distilled water

[4]Phosphate salts might contain chemically-bound water (v.g., monohydrate, dihydrate, etc.). You must factor in water molecule number in the formula weight.

NOTES

Questionnaire
Lab 3. Preparation of a Phosphate Buffer

1. Your lab manual contains a deduction of the Henderson–Hasselbalch equation using the dissociation of a weak acid, HA, to its conjugate base (A^-) and a H^+. Use the dissociation of a weak base to estimate *pH (yes, pH) of a weak base* in solution using the parameter pK_b.

$$BOH \leftrightarrows B^+ + OH^-$$

   where BOH = a weak base. Start by describing the degree of dissociation of BOH into OH^- and B^+ using K_b. Hint: $pOH = -\log [OH^-]$
2. Describe, stepwise, how you would prepare 0.5 L of a 0.15 M phosphate buffer, pH 12.5, using $K_3PO_4 \cdot H_2O$ (f.w. = 230.28) and K_2HPO_4 (f.w. = 174.18). Assume that after dissolving the salts your pH meter reading is 11.9.
3. What is the enzyme that catalyzes the reaction $CO_2 + H_2O \rightleftharpoons H_2CO_3$ in human tissues, including blood?
4. Since this reaction takes place in the absence of the enzyme, what is the physiological advantage in having such enzyme in the blood? [Hint: find what the *turnover* number is for this enzyme.]
5. This enzyme is also found in chloroplasts. What do you think the role of this enzyme is in plants?

NOTES

NOTES

NOTES

Laboratory N° 4
Determination of Protein Concentration Using the Bradford Method

Introduction

Proteins are the most diverse and abundant group of macromolecules within cells. Their functions include catalysis and regulation of numerous cellular metabolic processes (i.e., enzymes and signaling components, respectively) as well as forming structural components (e.g., the cytoskeleton and microtubules, etc.). Among the most useful parameters used in the study and characterization of proteins are their molecular weight and their concentration in the tissue or culture of interest. In this laboratory, we will determine the concentration of proteins using a spectrophotometric method. Protein concentration in cells or tissues is usually determined on a solution derived from breaking down the source cells followed by serial dilutions of the extract in an appropriate buffer or, for extracellular enzymes, dilutions of the growth medium or the perfusion fluid. Determination of protein concentration aids in characterizing enzymatic activity, which can be expressed as the amount of enzyme that catalyzes the transformation of one μmole of substrate to product per minute. Activity per mg protein is a parameter called *specific activity*. For non-enzyme proteins, measuring concentration will indicate their abundance and their possible role(s) in the physiology of the cell.

Spectrophotometry

The equation known as the Lambert–Beer law[5] describes the absorbance of a substance or a mixture in solution:

$$A = \varepsilon l c$$

where:

ε = Extinction coefficient, which is more accurately termed the absorption coefficient.

This is a proportionality constant which defines the efficiency or extent of absorbance by the compound. For macromolecules, such as proteins, DNA and RNA, ε is used in the form molar absorption coefficient, ε, which can be defined as the absorbance of a 1 M solution of the molecule at a given wavelength (λ) through a 1 cm path length. The units of ε are M^{-1} cm^{-1}. There are other forms of ε, such as $\varepsilon^{\%}_{cm}$.

l = Path length in cm. The path length is the thickness of the sample through which the light beam passes, and, for standard equipment, it is 1 cm.

A = Absorbance, which, despite what some industry standards say, has no units:

$[A] = (\cancel{M^{-1}} \cdot \cancel{cm^{-1}}) \cdot \cancel{M} \cdot \cancel{cm}$

Instrumentation

The measurement of light absorbance is usually obtained in an instrument called a spectrophotometer. The instrument consists of four major components (cf. Figure 9):

(1) Light sources, usually a tungsten visible wavelength lamp (VIS, 350–800 nm) and a deuterium, ultraviolet wavelength lamp (UV, 180–350 nm). **Please turn off deuterium lamp when not in use!**

---

[5] Although widely known as the Lambert–Beer law, based on the principle of priority it should be called the Bouguer–Beer–Lambert law.

DOI: 10.1201/9788743813224-4

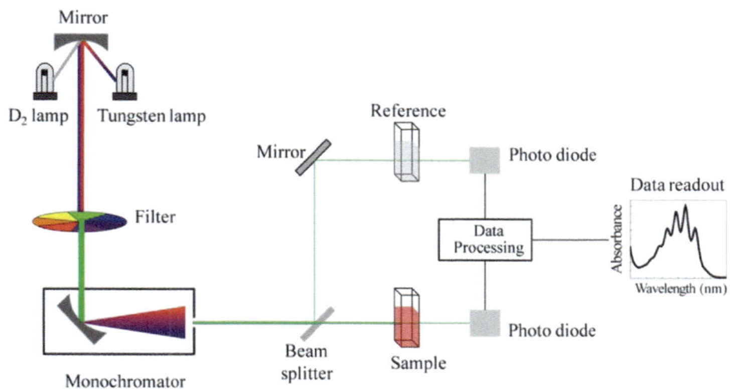

Figure 9. Diagram of a typical UV/Vis spectrophotometer (Wikimedia Commons).

(2) A monochromator that can be adjusted to allow only a single wavelength of electromagnetic energy to exit the light source.

(3) A sample holder to keep the sample between the light source and the photodetector.

(4) Two photodetectors that quantify the amount of light that is transmitted through the sample from the light source (Figure 9).

One photodetector is positioned behind the sample. This detector quantifies the amount of light that passes through the sample. The second detector, called the reference detector, quantifies the amount of light that is incident upon the sample. Both detectors work simultaneously because the light beam is split into two parts by a half-mirror. Therefore, the absorbance is a ratio of the light detected by the sample detector versus that detected by the reference detector. The photodetectors are attached to a digital or meter readout. In addition to these main components, the instrument may contain additional optical systems (e.g. lenses, slits, filters) to focus the light beam as it passes into the sample or into the photodetector.

During a spectrophotometric reading, the sample is contained in a small rectangular tube called a cuvette, which is inserted in the sample holder. For readings in the visible range, the cuvette is made of glass or polystyrene. Readings in the UV range require cuvettes that are made of quartz, because glass and polystyrene absorb in the UV.

Examples

1. The absorbance of a solution of the amino acid tyrosine ($\lambda = 280$ nm) is 0.75. The E_{280} for the amino acid tyrosine is 1500 M^{-1} cm^{-1}. The path length of the cuvette is 1 cm. What is the concentration of the tyrosine solution?

   Solution: $A = Elc \therefore c = A/El$

   Substituting $l = 1$ cm and $E = 1500\ M^{-1}cm^{-1}$, $c = 0.75/(1500\ M^{-1}cm^{-1})(1\ cm) = 5 \times 10^{-4}$ M.

2. $E_{470} = 108{,}427\ M^{-1}\ cm^{-1}$ for β-carotene in hexane. If A_{470} of a 1:10 dilution of a hexane extract from carrots is 0.432, what is the concentration of β-carotene in the original extract (in mg/mL)? Light path length = 1 cm, spectrophotometer is zeroed with hexane, β-carotene f.w. = 536.87 g/mol; measurements are made in glass cuvettes.

   Solution: as above, $c = 0.432/(108{,}427\ M^{-1}\ cm^{-1})(1\ cm) = 3.98 \times 10^{-6}$ M (1:100 dilution)= 3.98×10^{-4} M in the original hexane extract.

NOTES

Since 1 M of β-carotene contains 536.87 g/L solvent, 3.98×10^{-4} M = (536.87 g)(3.98×10^{-4} g/L) = 2.14×10^{-1} g/L = 2.14×10^{-1} mg/mL.

Spectrophotometric Determination of Protein Concentration

Most proteins do not display any color when seen under visible light. If a protein has a large content of aromatic amino acids, its concentration can be measured by UV absorbance, specifically A_{280}.[6] However, this method is not accurate if the proteins of interest do not have aromatic amino acids. The most common method for determining the concentration of proteins in solution relies on developing a colored complex between a protein and a dye. The protein–dye complex will absorb light at specific wavelengths. This absorption is directly proportional to the concentration of the protein and therefore measurable with an apparatus such as the spectrophotometer.

The method that we will use to determine protein concentration is Bradford's protein dye-binding assay. The dye Coomassie Brilliant Blue G-250 is a relatively hydrophobic substance that binds avidly to the hydrophobic regions of proteins. In dilute acid solution, the color of the Coomassie blue dye changes proportionally as it binds to protein, within a certain concentration range. As a result, the absorption maximum (λ_{max}) of the dye shifts from 465 nm to 595 nm, a bathochromic effect. This change in absorption can be used to quantify indirectly the amount of protein in the solution using a spectrophotometer.

Assays that involve color changes due to the interaction of a protein with a reagent are called colorimetric or chromogenic assays. The Bradford assay offers two major advantages, one being the ease of performance and the other that very few substances interfere with it. Some of the interfering substances are high concentrations of detergents such as Triton X-100 and sodium lauryl sulfate (SDS) which disrupt the binding of the dye to protein due to their amphipathic nature. The sensitivity of the Bradford assay is in the range of 1 µg to 100 µg of protein.

The Standard Curve

Standard curves provide a reference for measuring the amount of a substance in a solution of unknown concentration. In this particular lab, the substance is a protein. The curve is constructed by measuring the absorption of several solutions with known concentrations of protein in the range 10–100 µg[7]. A reaction between a dye and the protein renders a colored complex (in this case blue). The intensity of color is proportional to the amount of the dye–protein complex. The absorbance at 595 nm of solution(s) of unknown concentration can be compared to the standard curve. The concentration of the unknown must be in the range of the standard curve, in this case somewhere between 10 and 100 µg. If not, dilutions must be made. A fail-safe approach is to prepare a *decimal* serial dilution series: 1:10, 1:100, 1:1000, 1:10000, ..., $1:10^n$ (this can be expressed as $10^{-1}, 10^{-2}, 10^{-3}, 10^{-4}, ..., 10^{-n}$).

To determine the concentration of protein in your preparations, you will first construct a standard curve using a solution of known concentration of bovine serum albumin (BSA). Later, using serial dilutions, the unknown protein concentration of mitochondria will be determined by interpolation in the standard curve.

Procedure

1. Obtain a solution of protein whose concentration is known. BSA has been prepared in 0.15 M NaCl at a concentration of 1.0 mg/mL (recall that 1 mg = 1000 µg).

---

[6] Whereas concentration of proteins rich in aromatic amino acids can be determined by A_{280}, unstained nucleic acid concentrations are estimated by their A_{260}.

[7] Some dyes such as the bicinchoninic acid or alternate protocols for the Bradford's method may render standard curves and protein readings in the ng to mg range.

2. Label a set of test tubes (13 × 100 mm) as **B** and **1** to **5** using a black Sharpie marker.
3. Construct a table in your notebook (see example in Table 5).

Table 5

BSA standard curve					
Tube N°	1 mg/mL BSA (μL)	0.15 M NaCl (μL)	Bradford reagent (mL)	Protein (μg/tube)	A_{595}
B	0	50	2.5		
1	10	40	2.5		
2	20	30	2.5		
3	30	20	2.5		
4	40	10	2.5		
5	50	0	2.5		

4. Following the table, fill the provided glass tubes sequentially with BSA/0.15 M NaCl dilutions and the Bradford reagent, which contains a dilution of the dye Coomassie Brilliant Blue G-250.[8]
5. You may use small squares of Parafilm to facilitate mixing.
6. Full color development should occur in 5 minutes.
7. Calculate the amount of protein each tube contains and enter in the column headed Protein (μg/tube).
8. Quantify color development using the spectrophotometer. Assay the tubes within 1 hour. Use tube **B** (the blank) to set A_{595} to zero.
9. Measure A_{595} tubes 1 to 5. Write the values in the table in your notebook.
10. Plot your data on graph paper with μg protein on the abscissa and A_{595} on the ordinate. Strive to trace your standard curve within as many points as possible.[9]

Assay of Unknown Samples

In this portion of the laboratory, you will determine the concentration of protein in previously made mitochondria and wheat germ extract preparations (you may also use another sample if available). The idea is to prepare you to measure your own samples in subsequent labs and in your future employment or career path.

Dealing with unknowns and obtaining as much information about them as possible is one of the things that makes science so much fun. However, the enjoyment factor is nearly canceled out by the frustration level if the investigation is not carried out in an orderly fashion.

For example, the way to ascertain the concentration of protein in the unknown is to compare its absorbance to the standard curve created in the first part of this exercise. You might guess that a five-fold or a ten-fold dilution might do the trick. However, if the A_{595} does not fit in the curve, it will be necessary to dilute the sample again. If the absorbance reading still does not fit, you would have to try yet another dilution. This "shot in the dark" technique not only takes a lot of time, but it also uses up quite a bit of sample and provides plenty of opportunity for error. The most efficient way to determine the concentration of an unknown is by using serial dilutions. For example, by diluting the unknown 10-fold each time, you will be guaranteed to have one tube with an absorbance value that fits on the standard curve.

[8]**Note:** It is important to mix the tubes rapidly and thoroughly immediately after the dye is added, one tube at a time. Because the Bradford reagent contains phosphoric acid, avoid contact with the mouth or skin.

[9]If you do not obtain a linear graph, repeat the assay being more careful with your pipetting technique.

Once the concentration of unknown is determined for a particular tube, it is only a matter of a simple "back calculation" (using a dilution factor) to determine the amount of protein in the original sample.

Procedure

(C.2.1) Preparation of serial (decimal) dilutions

1. Obtain 100 μL of a protein sample prepared by your instructor. You may use, if available, potato tuber mitochondria and wheat germ preparations.
2. Pipette into properly labeled microfuge tubes (e.g. "M1," "WG1," see Table 5).
3. Label two sets of microfuge tubes as M2, M3, M4 and WG2, WG3, WG4.
4. Pipet 450 μL of 0.15 M NaCl into each tube. Use a P1000.
5. Add 50 μL of mitochondria to tube M2 (do not discard the remaining 50 μL of mitochondria prep in tube M1). Your volume in the tube should be 500 μL.
6. Add 50 μL of wheat germ preparation to tube WG2. Your volume in the tube should be 500 μL.
7. Tubes M2 and WG2 now contain a 10-fold dilution of the unknown, i.e. an order of magnitude less protein than the original tube.
8. Mix the tube gently by pipetting up and down; try to avoid creating bubbles. This action also rinses the walls of the pipet so that any solution remaining in the pipet will be of the same protein concentration as the tube.
9. Transfer 50 μL from tube M2 to tube M3 and from WG2 to WG3; mix gently. Tubes M3 and WG3 now contain a 10-fold dilution of the protein contained in tubes M2 and WG2, or 100-fold less protein than the original.
10. Now transfer 50 μL from tube M3 to M4 and from WG3 to WG4, mix gently.How many orders of magnitude less than the original do you have in these last tubes? ______.
11. Use a table like Table 5 and write these dilutions it as a fraction (___:_____) and also expressed as powers of 10 (_____$\times 10^{-n}$).
12. What volume of solution would you have ended up with if you had made this dilution directly from the original 25-μL sample? ____________

    [In other words, starting with 25 μL of unknown, how much 0.15 M NaCl would be necessary to achieve this same degree of dilution *without* using the serial dilution technique?]

(C.2.2) Color development of the unknowns

1. Label a set of glass tubes as M1 to M4 and WG1 to WG4. Transfer 50 μl of each of the dilutions to the corresponding tubes.
2. Construct a table in your notebook *similar* to Table 6 for generating the standard curve. Table 5 is only an *example*. You may have different extracts or more than just these two samples.

NOTES

Table 6

Sample source[10]	**Tube N°**	**Dilution**		**Bradford reagent (mL)**	**A_{595}**	**Dilution factor**	**Protein (µg/tube)**
		Inverse	Powers of 10				
Example: Wheat germ	WG1	—	10^0	2.5		1	
	WG2	1:10	10^{-1}	2.5		10	
	WG3	1:100	10^{-2}	2.5		100	
	WG4	1:1000	10^{-3}	2.5		1000	
Example: Potato tuber mitochondria	M1	—	10^0	2.5		1	
	M2	1:10	10^{-1}	2.5		10	
	M3	1:100	10^{-2}	2.5		100	
	M4	1:1000	10^{-3}	2.5		1000	

3. Add 2.5 ml of the Bradford reagent to each tube. Wait at least 5 minutes and then measure A_{595}.
4. Select the A_{595} value that falls in the middle region of the standard curve. This is the sample that falls within the useful range of the standard curve.
5. Use this A_{595} value for the back-calculation to determine the concentration of BSA in your original unknown solution.

Example

Let's say that by interpolation (or using an Excel trend equation) on the standard curve, the A_{595} of your unknown tube WG3 corresponds to an amount of 35 µg/tube. This should be transformed to concentration in µg/mL.
Since the standard BSA tubes and the unknowns had a total volume of 2.55 mL we have:

$$[\text{Total protein}] = (35\ \mu\text{g}/2.55\ \text{mL}) = 13.72\ \mu\text{g/mL}.$$

The protein concentration of the original unknown sample is obtained by multiplying the concentration of protein in tube by its dilution factor, reported in µg/mL and converted to mg/mL:

$$(13.72\ \mu\text{g/mL})(100) = 1,372\ \mu\text{g/mL} = \boxed{1.372\ \text{mg/mL}}$$

[10]These examples refer to exercises in this lab manual (wheat germ from Lab 5, potato tuber from Lab 6), but other protein sources can be used, for example, a macerated plant or meat portion, milk, egg white, yogurt, etc.

REPORT
Lab 4. Bradford Method

1. When preparing the protein standards, we used a tube (T) without protein, but including Bradford reagent and an equivalent volume of 0.15 M NaCl instead of protein. Can we just use plain *water* to prepare tube T instead?

   ❑ Yes ❑ No Explain.

2. The objective of this exercise was to train you in determining protein concentration in different extracts from biological samples. The protein concentration in a particular sample is used to standardize the activity of a protein such as an enzyme, a transporter, or receptor. Define the term "specific activity" in regard to an enzymatically catalyzed reaction:
3. Report your protein concentration for *each preparation* (i.e., your "best" determination) in the following table:

Sample source	**Tube N°**	**Dilution factor used for best determination**	**Protein (µg/mL)**
Wheat germ			
Mitochondria			

4. Why did you choose those particular dilution factors to get your "best" determination?
5. Include a print-out of your protein standard curve with your report. Indicate how can you estimate protein concentration of an unknown sample using the $\Delta \mathbf{A}_{595}/\Delta \mathbf{c}$ slope. Remember: **c** in this case is the absorbance of the sample *before* multiplying by the dilution factor.

NOTES

Use this sheet after you have completed labs 5 and 6. You will have your own samples from tissue extracts prepared for these exercises and will be able to measure protein after the fact.

Acid Phosphatase Specific Activity (Wheat Germ Extract)
Preparation dilution that falls in standard curve linear region: ☐ 10 ☐ 100 ☐ 1000 A_{595} of this unknown dilution: ________ ("Raw A_{595}") Corrected A_{595} = (Raw A_{595} X Dil. factor) (1 mg/1000 µg) µg/mL = ________ (10^{-3} mg/mL) = _____ mg/mL Phosphatase activity (µM/s) from Lab 5: ______, [Wheat germ total protein] = ____mg/mL **Acid phosphatase specific activity = (______ units/mL)/(_____ mg/mL)** **= ________ units/mg protein**

Cyt c Oxidase Specific Activity (Mitochondrial Sample)
Preparation dilution that falls in standard curve linear region: ☐ 10 ☐ 100☐ 1000 A_{595} of this unknown dilution: _________ ("Raw A_{595}") Corrected A_{595} = (Raw A_{595} X Dil. factor) (1 mg/1000 µg) µg/mL = ________ (10^{-3} mg/mL) = _____ mg/mL COX activity (units/mL) from Lab 6: ______, [Mitochondria total protein] = ____ mg/mL **COX Specific activity = (______ units/mL)/(_____ mg/mL) = ________ units/mg protein**

If you measured other samples, you may report them using a format similar to the ones above.

NOTES

NOTES

Laboratory N° 5
Enzyme Kinetics

Introduction: Part 1. Enzymes

Chemical reactions, including those that take place in biological systems, would proceed at very slow rates under standard temperature and pressure (STP) conditions. An increase in reaction rates could be achieved by raising the temperature of the environment, lowering the reaction's *activation energy*, or both. For most living organisms, heat is not a viable option to speed up biochemical reactions. Biological systems have evolved organic catalysts, known as enzymes, without which biochemical reaction rates would be too slow to maintain metabolism and growth processes. The great majority of known enzymes are *proteins* but a few of them, the ribozymes, depend on *RNA* for their catalytic activity.
Like other catalysts, enzymes lower the activation energy of reactions, they are not consumed themselves during their catalytic activity, and they are required in very small amounts. Enzymes differ from inorganic catalysts in several ways, among them: with some exceptions, enzymes are denatured by heat, they work optimally within a narrow range of pH and ionic strength, and, very importantly, they catalyze very specific reactions, allowing for the coexistence of a wide variety of catabolic and anabolic pathways in living organisms.

Definition of Enzymes

The study of enzymes and their activities is indispensable in understanding the molecular mechanisms of biological systems (Segel, 1975). An enzyme-catalyzed reaction in which a molecule (the substrate) is modified into a different one (the product) can be described as follows:

$$\mathrm{E} + \mathrm{S} \underset{k_{-1}}{\overset{k_1}{\rightleftharpoons}} \mathrm{ES} \overset{k_\mathrm{p}}{\rightarrow} \mathrm{E} + \mathrm{P}$$

where:
E = enzyme
S = substrate
ES = enzyme-substrate complex
P = product of the reaction
k = rate constants that describe the rate at which each of the reactions proceeds, as indicated by the arrows.
The *substrate* is the molecule that is acted on by the enzyme. In general, enzymes exhibit a high degree of *substrate specificity*. In other words, they bind to a single type of substrate and catalyze a single chemical reaction. Enzymes are usually named according to their substrate or the reaction they catalyze and the suffix *-ase*. For example, an enzyme that hydrolyzes proteins into smaller polypeptide chains is called a protease. An enzyme that catalyzes the synthesis of ATP is called an ATP synthase.
The first step in enzyme-mediated catalysis is the binding of the enzyme to its substrate to form an *enzyme–substrate complex*. This interaction is responsible for the reaction's specificity. Only substrates that "fit" into the *binding site* of the enzyme will undergo catalytic conversion to the product (cf. Figure 10). In many cases, the actual binding of the substrate to the enzyme distorts the spatial conformation of both enzyme and substrate in such a way that the formation of the product is facilitated. In the most common mechanism, the binding of substrate to the binding site brings certain chemical bonds of the substrate in close proximity to the lateral groups of certain amino acids of an enzyme.

DOI: 10.1201/9788743813224-5

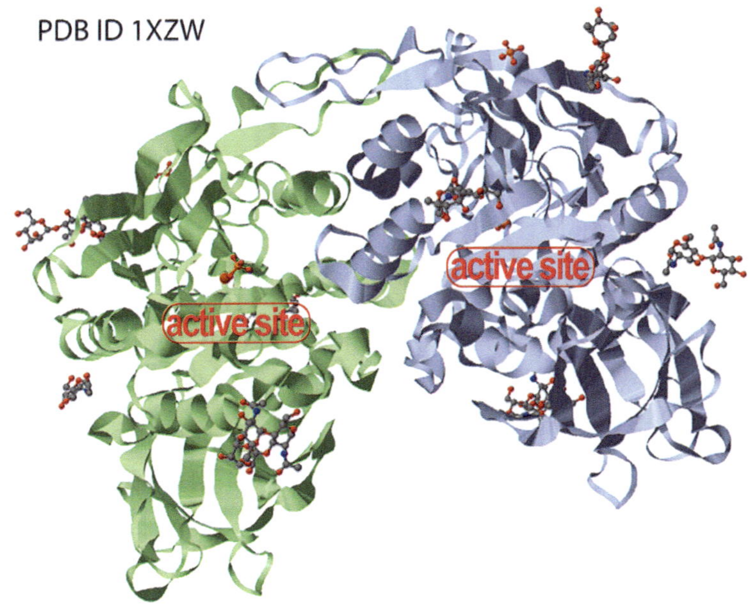

Figure 10. Structural model of the purple sweet potato acid phosphatase dimer, adapted from the Protein Database (www.rcsb.org/structure/1XZW).

The chemical microenvironment formed by the R groups facilitates the chemical reaction resulting in the conversion of substrate to product. The region of the enzyme that contains the lateral groups that bind the substrate and facilitates the chemical reaction is called the *active site* (Figure 10) of the enzyme. The lateral groups involved in the reaction are called *functional groups*. The formation of the ES complex is what lowers the activation energy of the reaction, allowing it to proceed toward the formation of the product.

Upon conversion of the substrate to the products, the enzyme and product(s) dissociate from one another, and the enzyme is free to interact with another substrate molecule. This property of enzymes makes them very efficient in catalysis. As a result, very few molecules of enzyme are required to convert large quantities of substrate to product.

Figure 10 shows a structural model of the well-characterized acid phosphatase from purple sweet potato (Schenck, 2005). The enzyme consists of a dimer, and the figure highlights the approximate locations of its active sites. Although we are studying a wheat enzyme, it is assumed that the acid-base catalytic mechanism is like that of the purple sweet potato enzyme. Phosphatases catalyze reactions opposite to those of phosphorylases, i.e., removing phosphate groups from a wide range of molecules, usually resulting in their deactivation.

Activity

The rate of enzyme reaction is referred to as the *velocity* of the reaction. It is most common to express the velocity in *amount of product formed per minute*. The *initial reaction velocity*, $\mathbf{v}_0$, of an enzymatic reaction varies with the substrate concentration, [S]. This can be illustrated by plotting v_0 vs. [S] (Figure 11).

The resulting curve is known as the Michaelis–Menten plot, which was named for Leonor Michaelis and Maud Menten,[11] two pioneers in the field of enzyme kinetics analysis. From this plot, an equation can be derived to describe the characteristics of an enzyme-catalyzed reaction. The equation is called the *Michaelis–Menten equation*. The common form of the equation is:

[11] Do not be misled by this first name. Dr. Leonor Michaelis was a man, while Dr. Maud Menten was a woman.

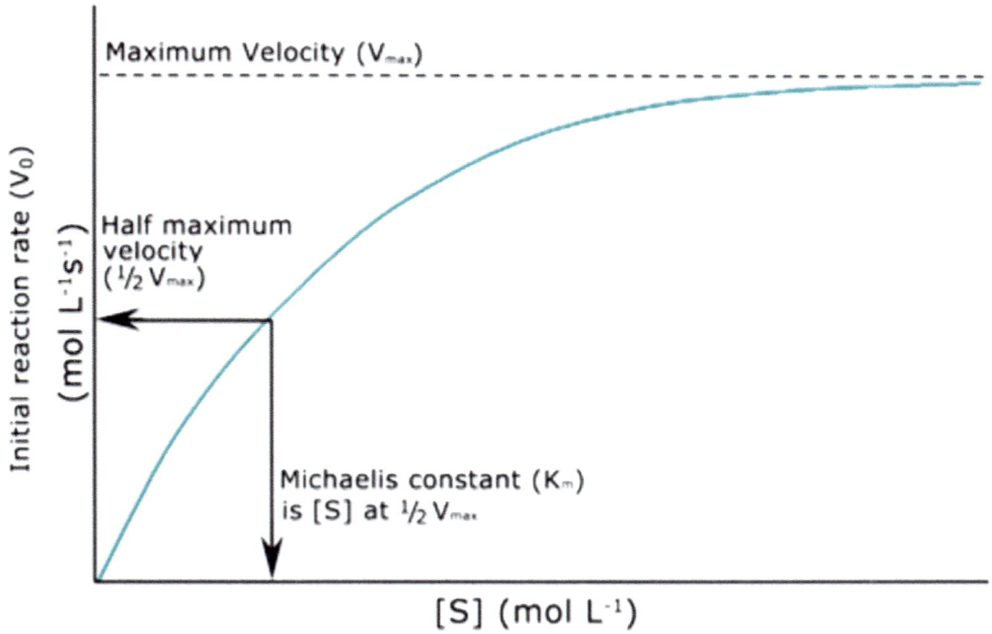

Figure 11. The Michaelis–Menten plot for an enzyme-catalyzed reaction as a function of initial velocity on substrate concentration.

$$v_0 = \frac{V_{max}[S]}{K_m + [S]} \tag{12}$$

where:
v_0 = initial reaction velocity, expressed as amount of product produced per unit time, i.e. mol L^{-1} s^{-1}
V_{max} = maximal reaction velocity; achieved when all the enzyme active sites are occupied with substrate molecules. This condition is called substrate saturation.
[S] = substrate concentration (mol L^{-1})
K_m = Michaelis constant = $(k_{-1} + k_p)/k_1$
Two important parameters, V_{max} and K_m, can be determined graphically from the *Michaelis–Menten plot* (Figure 11). V_{max} is the *maximal rate* at which an enzyme catalyzes a reaction. It is expressed as the amount of product formed per unit time. The units of V_{max} are the same as for v_0. This value can be extrapolated from the asymptotic plateau of the curve on the Michaelis–Menten plot. The plateau region represents reaction conditions in which the concentration of substrate far exceeds the concentration of enzyme active sites. Therefore, all active sites will be occupied at any given time. Under these conditions, the enzyme is said to be saturated with the substrate. The V_{max} is not a constant because it depends on the amount of enzyme in the reaction, but it is a very useful number because it can be used to calculate a constant called the *turnover number*, k_p, the number of substrate molecules transformed to product by one enzyme molecule per unit time. It can be calculated from V_{max} using the equation:

$$k_p = \frac{V_{max}}{[E_t]} \tag{13}$$

where [Et] = the total concentration of enzyme in the reaction with units of µg/mL or molarity (M), if molecular weight of the enzyme is known.
The turnover number is useful because it is a measure of the efficiency of the enzyme. The higher the k_p, the more effective the enzyme.

NOTES

The K_m or *Michaelis constant* is the other very useful parameter that is derived from the Michaelis–Menten equation. K_m is related to the rate constants of the reaction. In simple terms, it is the ratio of the rate at which ES forms (k_1) to the rate at which ES dissociates ($k_{-1} + k_p$). The dissociation of ES is the sum of the rate that the substrate dissociates from the enzyme before it undergoes the enzyme catalyzed reaction (k_{-1}) and the dissociation of product following the reaction (k_p).
There are three important points to remember about K_m:

1. The K_m is a measure of the apparent affinity of the substrate for the enzyme. In other words, it indicates how well the substrate fits into the binding site of the enzyme. The substrate with the lowest K_m value has the highest apparent affinity for the enzyme. The "best" substrate is that which has the highest V_{max}/K_m.
2. K_m is directly related to V_{max} as it represents the substrate concentration at half the maximal velocity (½ V_{max}) of the enzyme.
3. The K_m of an enzyme is a constant for every individual enzyme with a particular substrate. Therefore, it serves as a fingerprint for the presence of a specific enzyme. This is very useful for investigators who are determining whether an enzyme is present or absent in a particular cell or tissue or to determine the affinity for similar substrates.

   The units of K_m are molarity (M).

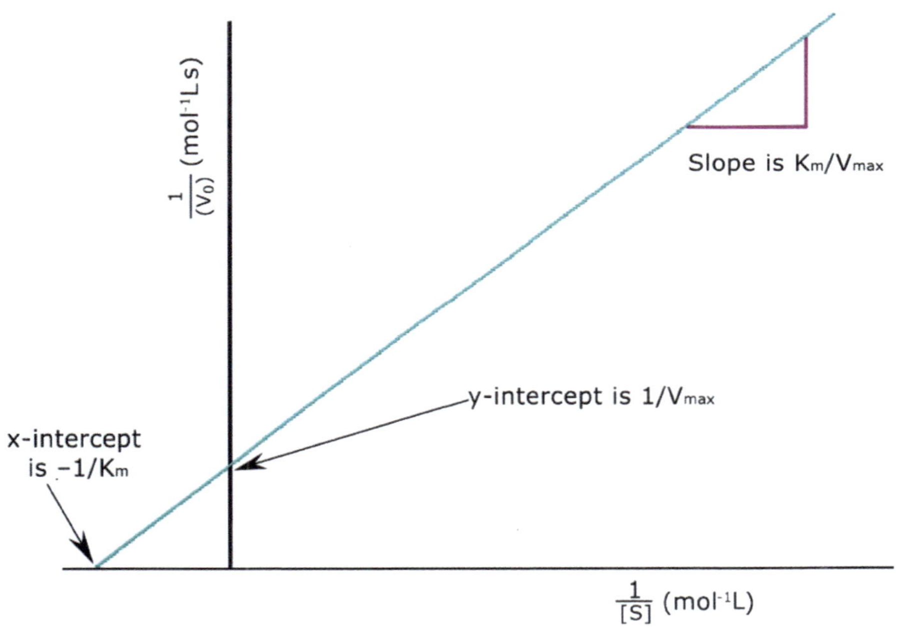

Figure 12. The Michaelis–Menten plot for an enzyme-catalyzed reaction as a function of initial velocity on substrate concentration.

The fact that the Michaelis–Menten plot is a rectangular hyperbola makes it difficult to accurately determine the V_{max} and K_m values from a graph. Therefore, the terms in Equation (2) are often rearranged and used to generate a plot that gives a straight line. This equation is called the *Lineweaver–Burk equation*:

$$\frac{1}{v_0} = \frac{K_m}{V_{max}} \cdot \frac{1}{[S]} + \frac{1}{V_{max}} \tag{14}$$

This equation is in the form y = mx + b and the plot $1/v_0$ versus $1/[S]$ is a straight line (Figure 12). The intercept on the $1/v_0$ axis is $1/V_{max}$ and the intercept on the $1/[S]$ axis is $-1/K_m$. The slope of the line on the plot is K_m/V_{max}.

NOTES

Experimental Procedures

In this laboratory exercise you will analyze the activity and kinetic properties of an enzyme derived from wheat germ. The wheat kernel, or caryopsis, consists of three parts: (1) the embryo or germ that will develop into the new plant, (2) the starchy endosperm and the aleurone which supply the embryo with nutrition, and (3) the hull or bran which protects the grains. The wheat germ also is a common source of enzymes and nucleic acids for studies in biochemistry and cell biology. You will analyze the enzyme *acid phosphatase* from wheat germ.
Acid phosphatase catalyzes the hydrolysis of phosphate groups from proteins, nucleic acids and lipids that are stored in the seed. The phosphate is utilized by the growing embryo upon germination. To measure the enzyme activity, you will use the artificial enzyme substrate *p-nitrophenol phosphate*. This substance is colorless. However, after hydrolysis by acid phosphatase, it yields phosphate and nitrophenol, which is yellow (Figure 13). The amount of yellow color generated by the catalysis is a direct measure of the amount of nitrophenol produced and therefore is an indicator of the enzyme activity.

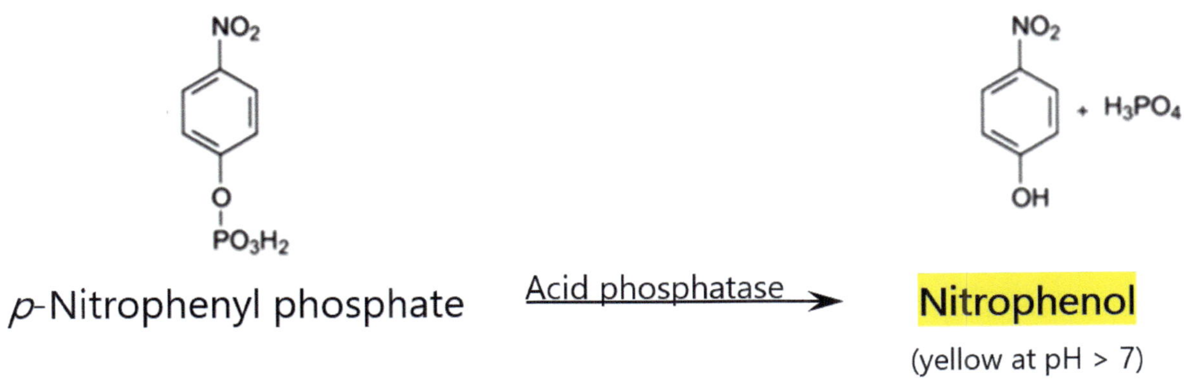

Figure 13. Dephosphorylation of p-nitrophenyl phosphate catalyzed by wheat germ acid phosphatase.

In the first exercise you will measure the velocity of the reaction catalyzed by acid phosphatase that is extracted from wheat germ. In addition, you will measure the velocity of commercially available purified acid phosphatase (part 2) as a standard for the enzyme's activity (part 3). By comparing the two values, you will determine the amount of enzyme that is present in the wheat germ extract.

(1) Preparation of a standard curve for the reaction product

- In order to quantify the amount of nitrophenol generated in your enzymatic assays, prepare six standards containing known concentrations of the product, nitrophenol. Place 1 mL of each of the standards into six cuvettes that have been labeled S1–S6.
- Read A_{410} of the tubes with the spectrophotometer. Use the S1 cuvette as your blank to set $A_{410} = 0$. Enter the data in a table such as Table 7.
- Construct a nitrophenol standard curve by plotting the A_{410} of tubes S1–S6 on the ordinate vs. the concentration of nitrophenol on the abscissa for each tube. You may plot your values in an Excel sheet (guided by your instructor) or create a millimetric grid as a PDF from https://inkpx.com/graph-paper/graph-paper-maker settings: Paper Size = USLetter, Orientation = Portrait, Margin (px) = 67, Grid Size = 1 mm, Line Weight (px) = 0.5, Line color: Choose; Show Major/Minor Lines) 1 inch/cm intervals); press Download

NOTES

Table 7

Nitrophenol standard curve for acid phosphatase assays

Tube	Nitrophenol (μM)	A_{410}	Tube	Nitrophenol (μM)	A_{410}
S1	0		S4	50	
S2	12.5		S5	100	
S3	25		S6	200	

(2) Preparation of acid phosphatase from wheat germ extract

The preparation of the wheat germ extract will be performed by the instructor at the beginning of class. Observe the procedure and record it in your notebook.

- Place 1 g of wheat germ into a mortar containing 30 mL of enzyme extraction buffer.
- Grind the tissue with a pestle until the mixture forms a homogeneous suspension.
- Transfer the solution to a centrifuge tube and spin for 15 min at 14,000 × g.
- Remove the supernatant to a clean tube. This is the wheat germ extract. Label this tube as *Extracted Acid Phosphatase.*

(3) Determination of acid phosphatase activity in a commercial preparation.

All steps in this assay are to be performed by students

- Obtain a tube rack containing 12 test tubes. Number six of the tubes **A1** through **A6** and six more as **B1** through **B6**.
- Place 0.5 mL of 1.5 % (w/v) KOH into each of the 12 tubes.
- Obtain two larger test tubes and label them **A** and **B**.
- Add 5 mL of phosphatase substrate solution to tube **A** and 5 mL to tube **B**. Tube **A** will contain the reaction catalyzed by purified acid phosphatase. Tube **B** will contain the reaction catalyzed by the extracted acid phosphatase.
- Transfer 0.5 mL of the substrate solution from tube **A** to tube **A1**. Transfer 0.5 mL of the substrate solution from tube **B** to tube **B1**. These two tubes are the 0 time values for each reaction.
- Place 50 μL of pure (commercial) acid phosphatase (125 μg at 2.5 μg/μL) into tube **A** and 0.2 mL of wheat germ extract into tube **B**.
- Mix the tubes gently and immediately start timing the reactions.
- At the times indicated below, remove 0.5 mL of the solutions in tubes **A** and **B** and place them into their corresponding tubes (**A1–A6** and **B1–B6**).[12]

2.5 min	**A2**, **B2**
5 min	**A3**, **B3**
10 min	**A4**, **B4**
15 min	**A5**, **B5**
20 min	**A6**, **B6**

[12] You will be instructed to use smaller glass tubes (e.g., 10 mm diam.) or appropriately sized spectrophotometer cuvettes.

Table 8

Acid phosphatase assay for commercial preparation (Series A) and wheat germ extract (Series B)

Tube	A_{410}	Nitrophenol produced (µM) = F·A_{410}
A1		
A2		
A3		
A4		
A5		
A6		

Tube	A_{410}	Nitrophenol produced (µM) = F·A_{410}
B1		
B2		
B3		
B4		
B5		
B6		

The KOH in the numbered tubes serves two functions. First, it stops the reaction because the acid phosphatase is inactive at alkaline pH (hence the name *acid* phosphatase). Second, the alkaline conditions turn the liberated nitrophenol to a more intense yellow (why?). This makes the assay more sensitive.

- Transfer the content of the tubes into cuvettes and read the absorbance of your assay tubes **A1–A6** and **B1–B6**. Enter the data into tables such as Table 8:
- Using the nitrophenol standard curve, determine the concentration of nitrophenol produced (µM) in each assay tube (A1–A6 and B1–B6) and record your data in your Table 8.
- Plot the concentration of nitrophenol produced (Y axis) against time (X axis) for sets A and B. Convert min to s. Strive to trace a straight line through the plotted points. You may find that the latest time points may not fall onto a straight line with respect to the earlier time points. If this is the case, draw the line through the first three time points on the plot. These three points should have a linear relationship. (Alternatively, you may do your plot in Excel and estimate the v_0 for the reactions through an equation.)
- From the graph, calculate the initial velocities (v_0) for the two reactions in µM/s.
- The v_0 that you calculate represents the V_{max} for the enzyme because the amount of *p*-nitrophenol phosphate added to each reaction was at a concentration that saturates the enzyme.
- Given the fact that you added 125 µg of purified acid phosphatase to tube **A** (or 50 µL of a 2.5 µg/µL stock), estimate the amount (µg) of acid phosphatase that was present in the 0.2 mL of wheat germ extract in tube **B**. **Hint**: To calculate the amount of enzyme in the wheat germ extract, you must first convert the V_{max} from the purified enzyme to a constant that can be applied to the enzyme in the wheat germ extract. This constant is the turnover number (k_p). which can be used to calculate the amount of enzyme in the wheat germ extract.

(4) Determination of V_{max} and K_m for acid phosphatase

In this part of the exercise, you will use the information on v_0 that you obtained above to determine the V_{max} and K_m of purified (commercial) acid phosphatase. This is to double-check the value for V_{max} that we assumed in the first part of the laboratory.

Choose a time point in the enzyme assay that lies within the linear range of the initial velocity. This is critical because enzyme catalytic rates decline with time due to depletion of the substrate, accumulation of the product, and enzyme inactivation. Hence, it is important that you choose a time that represents a true initial velocity for the reaction.

Second, you will set up a set of acid phosphatase assays containing increasing amounts of substrate, but a constant amount of enzyme. After incubating the assays for the predetermined time, you will determine the v_0 for each assay and prepare a Lineweaver–Burk plot for the enzyme. From this plot, you will determine the K_m and V_{max} for acid phosphatase.

- Look closely at the plot of µM nitrophenol vs. time generated above. Choose a time point at which the accumulation of product is still within the linear range of the assay, i.e., find a time-point that lies on the straight line. An incubation time of 5 min (300 s) might still be in the linear region of the curve. This time point will be the incubation time for the kinetic assay in this exercise.
- Seven solutions with different concentrations of *p*-nitrophenyl phosphate have been prepared for your use by the instructor. Label test tubes from **1** to **7** and transfer 1 mL of each substrate solution to a corresponding tube as given on the table below. These tubes represent assay mixtures with varying amounts of substrate.
- Add 50 µl of purified acid phosphatase (125 µg, i.e., if using a 5 µg/µL commercial preparation) to each tube at 30-second intervals. Start timing when you add the enzyme to the first tube.
- At the end of 2.5 min, add 1 mL of 1.5% (w/v) KOH to each tube at 30-second intervals to stop the reactions.

  Note: By staggering the addition of enzyme and KOH at 30-second intervals, you ensure that each reaction has proceeded for exactly 2.5 minutes.
- Quantify the concentration of nitrophenol produced (µM) in each reaction using the standard curve from part 2). Convert these values to v_0 values by dividing the concentration of nitrophenol produced by the reaction time. Enter all values in a table such as Table 9.

Table 9

Data set for estimation of V_{max} and K_m for purified acid phosphatase

Tube	A_{410}	Nitrophenol (µM)	v_0 (µM/s)	$1/v_0$ (s/µM)	[S] (µM)	1/[S] (μM^{-1})
1					0	
2					50	
3					100	
4					250	
5					500	
6					750	
7					1000	

- Prepare a Michaelis–Menten plot with the above data. Use an Excel sheet or download millimetric paper from https://inkpx.com/graph-paper/graph-paper-maker and proceed as described above.
- Estimate the K_m and V_{max} values from this plot. Express the values in the units of µM and µM/s, respectively.
- Prepare a Lineweaver–Burk plot of $1/v_0$ versus 1/[S].
- Determine the K_m and V_{max} from the Lineweaver–Burk plot. Are the values different from those that you estimated from the Michaelis–Menten plot?

NOTES

Questionnaire
Lab 5. Enzyme Kinetics

Part 1. Standard curve

1. Record your data for the nitrophenol standard curve in the table below:

Nitrophenol standard curve for acid phosphatase assays

Tube	Nitrophenol (μM)	A_{410}	Tube	Nitrophenol (μM)	A_{410}
S1	0		**S4**	50	
S2	12.5		**S5**	100	
S3	25		**S6**	200	

- Construct a nitrophenol standard curve by plotting the A_{410} of tubes S1–S6 on the ordinate vs. the concentration of nitrophenol on the abscissa for each tube. You may plot your values on an Excel sheet or download graph paper from https://inkpx.com/graph-paper/graph-paper-maker settings: Paper Size = USLetter, Orientation = Portrait, Margin (px) = 67, Grid Size = 1 mm, Line Weight (px) = 0.5, Line color: Choose; Show Major/Minor Lines)1 inch/cm intervals); press Download
- Obtain a curve in the form y = mx + b in which y = A_{410} and b = ordinate to the origin.
- In either case, determine the slope (m) and the ordinate to the origin (b).
- Establish a conversion factor (A_{410} to concentration) by obtaining 1/m:

$$m = \therefore F = 1/m = \_\_\_\_\_\_\_\_$$

- Read the absorbance of your assay tubes **A1–A6** and **B1–B6**. Enter the data into tables such as the ones below and convert A_{410} to nitrophenol produced (μM) in each assay tube (μM)

Part 2. Determination of initial velocity of acid phosphatase preparations (Table 10)

Table 10

Acid phosphatase assay for commercial preparation (Series A) and wheat germ extract (Series B)

Tube	A_{410}	Nitrophenol produced (μM) = $F \cdot A_{410}$	Tube	A_{410}	Nitrophenol produced (μM) = $F \cdot A_{410}$
A1			**B1**		
A2			**B2**		
A3			**B3**		
A4			**B4**		
A5			**B5**		
A6			**B6**		

- Plot the concentration of nitrophenol produced (ordinate) against time (abscissa) for sets A and B. Convert min to s.
- Aim to trace straight lines through the plotted points or you may want to use Excel and plot a trendline.

- You may find that the latest time points may not fall onto a straight line with respect to the earlier time points. If this is the case, draw the line through the first three or four time points on the plot. These points should have a linear relationship.
- If you use Excel to plot, estimate the v_0 for the reactions through a linear equation (restrict the data range to the first three or four points).
- From the graph, calculate the initial velocities (v_0) for the two reactions in µM/s:

Commercial preparation (A) ______ µM/s Wheat germ (B) ______ µM/s

- The v_0 that you calculate actually represents the V_{max} for the enzyme because the *p*-nitrophenol phosphate added to each reaction was added at a concentration that saturates the enzyme.
- If 125 µg of commercial acid phosphatase added to tube **A** (50 µL from a 2.5 µg/µL stock) produced ____ µg of nitrophenol, estimate the amount in µg of acid phosphatase in the 0.2 mL of wheat germ extract in tube **B** by comparing its acid phosphatase activity (_____ µg of nitrophenol, *equivalent* to the activity of _____ µg of purified enzyme).
- Given that you added 125 µg (in 50 µL from a 2.5 µg/µL stock) of the commercial prep of acid phosphatase to tube **A**, estimate the amount (µg) of acid phosphatase that was present in the 0.2 mL of wheat germ extract in tube **B**.
- Convert V_{max} to k_p (turnover number) for the commercial preparation:

$k_p = V_{max}/[E]_t$ = _____ µg/µL (= mg/mL); where $[E]_t = 2.5$ mg/mL

∴ for wheat germ extract: $[E]_t = V_{max}/k_p$ = _____ µg/µL (= mg/mL)

Part 3. Determination of V_{max} and K_m for acid phosphatase

✓ Last time point in the enzyme assay that lie within the linear range of the initial velocity: ______ min (= ______ sec).

- Prepare a Lineweaver–Burk plot for the enzyme. From this plot, you will determine the K_m and V_{max} for acid phosphatase. Enter all v_o values in a table such as the one given in Table 11.

Table 11

Data set for estimation of V_{max} and K_m for purified acid phosphatase

Tube	Nitrophenol (µM)	v_0 (µM/s)	$1/v_0$ (s/µM)	[S] (µM)	1/[S] (μM^{-1})
1				0	
2				50	
3				100	
4				250	
5				500	
6				750	
7				1000	

- Prepare a Michaelis–Menten plot with the above data using millimetric paper (downloadable from https://inkpx.com/graph-paper/graph-paper-maker or plot on Excel.
- Estimate the K_m and V_{max} values from this plot. Express the values in the units of µM and µM/s, respectively (Segel, 1975).

$$K_m = \_\_\_\_\_\_ \mu M \text{ and } V_{max} = \_\_\_\_\_\_ (\mu M/s)$$

- Prepare a Lineweaver–Burk plot of $1/v_0$ versus $1/[S]$.
- Determine the K_m and V_{max} from the Lineweaver–Burk plot.

$$-{K_m}^{-1} = \_\_\_\_\_\_ \mu M^{--1} \;\therefore\; K_m = \_\_\_\_\_\_ \mu M$$
$$\text{and } {V_{max}}^{-1} = \_\_\_\_\_\_ (\mu M/s)^{-1} \;\therefore\; {V_{max}}^{-1} = \_\_\_\_\_\_ (\mu M/s)$$

- How are these values different from those that you estimated from the Michaelis–Menten plot?

NOTES

NOTES

Laboratory N° 6
Assay of Cytochrome *c* Oxidase in Potato Tuber Mitochondria

Purpose

To study the last component of the electron transport system in mitochondria, namely the enzyme cytochrome *c* oxidase, which is involved in O_2 consumption in cellular respiration.

Introduction

The mitochondria originated through endosymbiosis from aerobic bacteria capable of using O_2 as the final electron acceptor in respiration. In eukaryotes, the Krebs cycle takes place in the mitochondria. The Krebs cycle provides donors to the electron transport chain in the mitochondrial inner membrane system, namely: $FADH_2$ (flavin adenine dinucleotide, reduced) and $NADH + H^+$ (nicotinamide adenine dinucleotide, reduced). Cytochrome *c* oxidase (EC. 1.9.3.1) is also known as COX or Complex IV of the electron transport system of the inner mitochondrial membrane and the bacterial plasma membrane. It accepts electrons from cytochrome *c* (cyt *c*) and transfers them to an acceptor oxygen molecule in the final step of the electron transport chain. Among COX inhibitors are CO and cyanide.

$$4\,\mathbf{Fe^{2+}\text{-cyt}\,\mathit{c}} + 8\,H^+ + O_2 \rightarrow 4\,\mathbf{Fe^{3+}\text{-cyt}\,\mathit{c}} + 2\,H_2O + 4\,H^+\uparrow_{[out]}$$

COX's activity accounts for more than 90% of the oxygen utilized by respiring organisms. The extrusion of protons from the mitochondrial matrix towards the intermembrane space allows for ATP synthesis as a pH gradient is formed across the inner mitochondrial membrane. Protons go through a $[H^+]$-ATP synthase when returning to the matrix before the gradient is dissipated (Figure 14). According to the chemiosmotic theory, a proton motive force is formed by the pH gradient and the electrochemical potential across the inner mitochondrial membrane, providing energy for ATP synthesis (Nelson and Cox, 2021).
The measurement of COX activity in vitro requires the use of glycosidic detergents to maintain its quaternary structure, a dimer in which each monomeric component contains thirteen apoproteins. In mammals, ten of the proteins that compose COX are encoded in the nucleus, whereas genes for the other three are in the mitochondrial DNA.

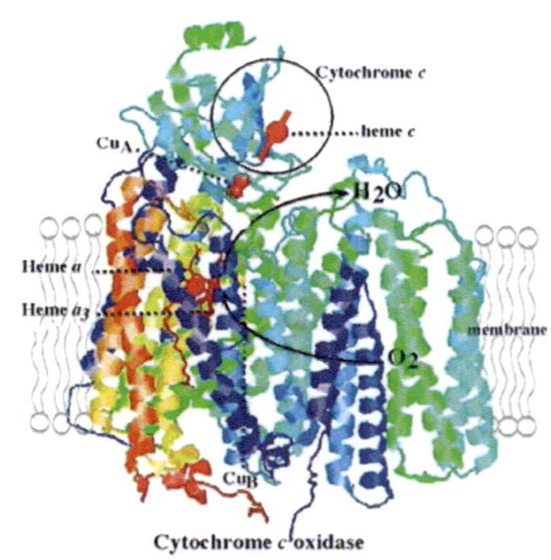

Figure 14. Mitochondrial complex IV. Adapted from Qin et al. (2007) and other works by S. Ferguson-.

Several maternally inherited diseases in humans are due to defects in the gene expression of the mitochondrial DNA (Brooker, 2020), including some that affect COX structure (e.g., assembly) or function. These diseases include fatal metabolic disorders such as Leigh syndrome, cardiomyopathy, Leber's hereditary optic neuropathy, and sensorineural deafness. Symptoms of these diseases start appearing in early childhood and chiefly affect tissues and organs with high-energy requirements (e.g., muscles, heart, brain) which have a high mitochondrial activity.
In this laboratory, we will use potato tubers as the source of mitochondria (Leaver et al., 1983) to assay the activity of the redox enzyme cytochrome *c* oxidase. We will be using cyt *c* as the substrate, since cyt *c* oxidation can be followed

DOI: 10.1201/9788743813224-6

spectrophotometrically by the oxidation (electron loss) of the heme iron in cyt *c*, which goes from Fe^{2+} (*ferrocytochrome*) to Fe^{3+} (*ferricytochrome*) and thereby causing a *decrease* in the A_{max} for ferrocytochrome at 550 nm (A_{550}).

Procedure

Part A. Isolation of mitochondria from potato (*Solanum tuberosum*) tubers

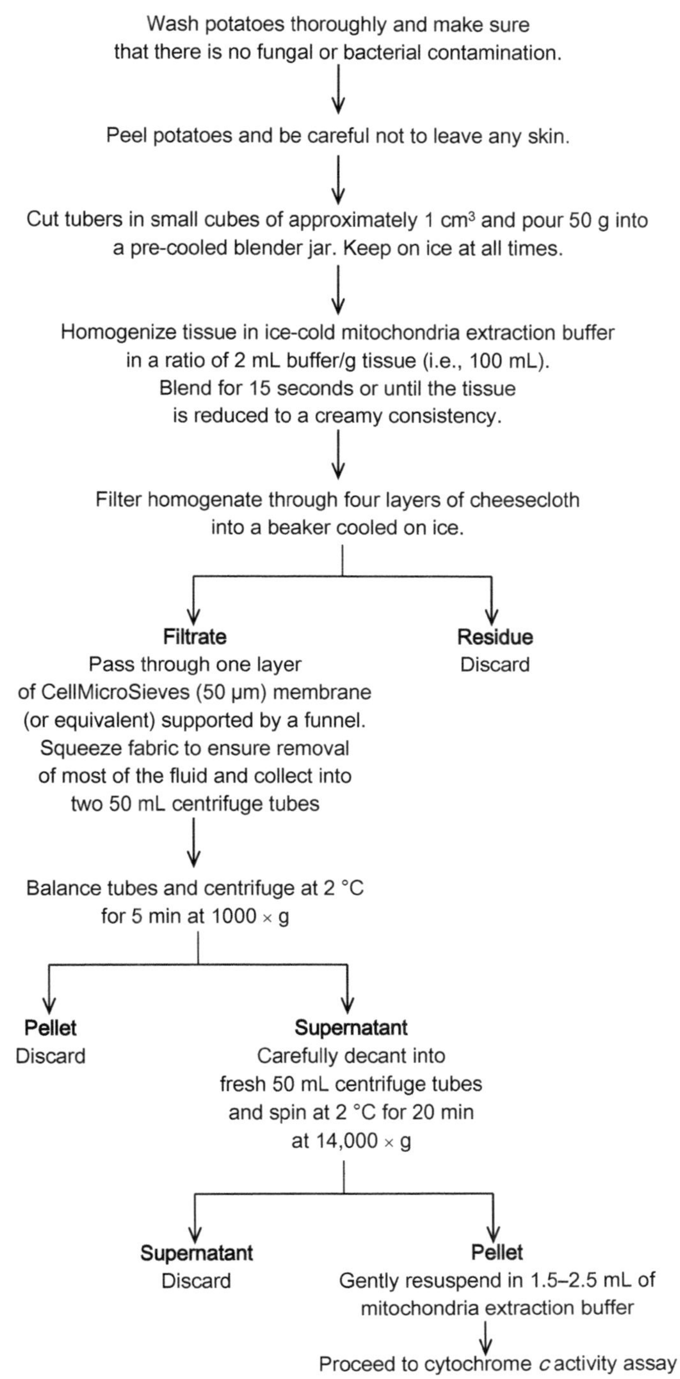

NOTES

Part B. Cytochrome *c* oxidase activity in *Solanum tuberosum* mitochondria

- Warm up the spectrophotometer's tungsten lamp and set the wavelength (λ) to 550 nm.
- Set up a reaction scheme according to Table 12.

Table 12

Sample	Assay buffer (μL)	Sample (μL)	Enzyme dilution buffer (μL)	Ferrocytochrome c substrate Solution (μL)
Reagents blank	950	0	100	50
Unknown sample 1	950	____ = x_1	$100 - x_1$ = ____	50
Unknown sample 2	950	____ = x_2	$100 - x_2$ = ____	50
Unknown sample 3	950	____ = x_3	$100 - x_3$ = ____	50
Unknown sample 3	950	100	0	50

Notice that x_i can represent increasing volumes of the mitochondrial prep (e.g. 50, 75 and 100 μL); therefore, the enzyme dilution buffer that needs to be added is $100 - x_1$ μL. For example, if you are instructed to use 75 μL of mitochondrial prep you must add 25 μL of enzyme dilution buffer.

1. Zero the spectrophotometer with 950 μL of 1× assay buffer.
2. Add a suitable volume of mitochondrial suspension to the cuvette and bring the reaction volume to 1050 μL with 1× enzyme dilution buffer and mix by inversion.
3. Start the reaction by the addition of 50 μL of ferrocytochrome *c* substrate solution and mix by inversion.
4. *Read the* A_{550} *immediately* (t_0) due to the rapid reaction rate of cyt *c* oxidase.
5. Keep taking readings every 15 seconds for 3–5 min. Background values are expected between 0.001 and 0.005 A_{550}/minute.
6. Estimate your ΔA_{550}*/minute* by dividing the ΔA_{550} by the total number of minutes you kept taking readings. $\Delta A_{550} = A_{550}t_f - A_{550}t_0$, where t_f = time at last reading (e.g., 3 min) and t_0 = zero time (initial reading).
7. Recall that 30 s = 0.5 min, 20 s = 0.333 min, 15 s = 0.25 min and so on.
8. Calculate the activity of the sample as follows:

$$\text{Units}/\text{mL} = \frac{(\Delta A_{550}/\text{min})\,(\text{dilution})(1.1)}{(21.84)(\text{vol. of enzyme})}$$

where:

- $\Delta A_{550}/\text{min} = A_{550}/\text{min}_{\text{(sample)}} - A_{550}/\text{min}_{\text{(blank)}}$
- dilution is the dilution *factor* of enzyme or sample[13]
- 1.1 refers to reaction volume in mL
- vol. of enzyme (mitochondria sample, i.e., 50, 75, 100 μL, etc.) should be converted to mL (i.e., 0.05, 0.075, 0.1 mL)
- 21.84 = $\Delta\varepsilon^{mM}$ between ferrocytochrome *c* and ferricytochrome *c* at 550 nm.

[13] If your mitochondrial preparation is not diluted, the dilution factor is 1.

NOTES

Preparation of Reagents

Mitochondria extraction buffer (200 mL per section)
(0.4 M mannitol, 1 mM EGTA, 25 mM MOPS, 0.1% (w/v) bovine albumin serum, 8 mM cysteine, 40 mM β-mercaptoethanol; pH 7.8 adjusted with KOH).

(5×) Assay buffer stock (10 mL per section)
(50 mM Tris-HCl, pH 7.0; 600 mM KCl)

(1×) Assay buffer stock
(10 mM Tris-HCl, pH 7.0; 120 mM KCl)
In a 15 mL Falcon tube, place 2.5 mL of (5×) assay buffer stock and add 10 mL of sterile diH_2O. Mix well and keep at room temperature.

(2×) Enzyme dilution buffer (10 mL per section)
(20 mM Tris-HCl, pH 7.0; 500 mM sucrose)

(1×) Enzyme dilution buffer
(10 mM Tris-HCl, pH 7.0; 250 mM sucrose)
In a 15 mL Falcon tube, place 5 mL of (2×) enzyme dilution buffer stock and add 5 mL of sterile diH_2O. Mix well and keep at room temperature.

0.1 M dithiothreitol for DTT working solution
Dissolve 200 μL of 1 M DTT stock (D7959) to 0.1 M by the addition of 0.8 mL of sterile diH_2O. Mix well, keep chilled. Return to freezer after use.

Ferrocytochrome c substrate solution (0.22 mM)

- Dissolve 2.7 mg of cytochrome *c* (MW 12,384 Da) in 1 mL of sterile diH_2O and add 5 μL of 0.1 M DTT solution.
- Mix gently and wait for 15 min. The color of the solution should go from dark-orange red to pale-purple red.
- Mix 50 μL of the cytochrome *c* solution prepared above and 950 μL of (1×) assay buffer, respectively. This renders a 1:20 dilution of cytochrome *c*.
- Using (1×) assay buffer as a blank to zero the spectrophotometer, measure A_{550} and A_{565} of the 1:20 cytochrome *c* dilution, and determine the A_{550}/A_{565} ratio:

$$A_{550} = \_\_\_\_\_ \quad A_{565} = \_\_\_\_\_ \quad A_{550}/A_{565} = \_\_\_\_\_$$

Note: The A_{550}/A_{565} ratio should be between 10 and 20. If the ratio is less than 10, the substrate has not been sufficiently reduced, and it should be left for an additional 15–20 min or prepared with a fresher 0.1 M DTT solution.

Enzyme sample
Best results are achieved when enzyme activity is between 0.4 and 4.0 milliunits of cytochrome *c* oxidase. Different amounts of enzyme preparation should be assayed to find a linear range for a particular sample.

NOTES

Questionnaire
Lab 6. Cytochrome c Oxidase Activity

Your lab discussion should include, but not necessarily be limited to the following questionnaire:

- ✓ Graph your reaction's A_{550} vs. time (s) i.e., your raw values. You may include the results obtained by other groups in your section. That way you will have activities from several dilutions of the mitochondrial preparation.
- ✓ Compare the raw values (from your graph) for all dilutions. Do their curves show the same slope?

❑ Yes ❑ No

  - If yes, is this expected?
  - If no, what could be the reason?

- ✓ Once the units/mL are calculated for all dilutions, are they the same?

❑ Yes ❑ No

  - If yes, is this expected?
  - If no, what could be the reason?

- ✓ From a subjective perspective, in your opinion, is this reaction slow or fast? Explain your answer.
- ✓ Why is cytochrome *c* oxidase one of the most important enzymes for aerobic organisms?
- ✓ Find examples of poisons whose target is cytochrome *c* oxidase (or, from a biochemical perspective, examples of cytochrome *c* oxidase inhibitors!).

NOTES

NOTES

Laboratory N° 7

Molecular Evolution: Separation of Fish Proteins Using Polyacrylamide Gel Electrophoresis

Introduction

In his classic book "The Origin of Species," Charles Darwin proposed that all living organisms are derived from a common ancestor through a process that he called "descent with modification," in which hereditary changes occur over many generations.

Hereditary changes are brought about by alterations in the DNA (genotype) called mutations (see exercise "*Molecular Biology of Sickle-cell Anemia*" in this manual), which result in the alteration of the structure of DNA-encoded proteins, leading to novel (phenotypic) traits. Thus, DNA provides both for the continuity of the traits through generations, but it also accounts for the variation that can lead to genetic diversity within populations.

At the genetic level, *evolution* can be defined as changes in a gene pool (collective genotype of a certain species) over time. Changes in the gene pool are produced by natural selection, a process that favors some individuals and not others, by giving them traits that may enhance, for example, their survival, reproductive rate, capacity for adaptation, or metabolic processes.

Evidence for genetic evolution comes from the knowledge of the nucleotide sequences in the DNA of many contemporary organisms that display a great deal of similarity in their genes. For example, the *hox* gene family has been shown to control the embryonic development in animals from groups as diverse as fruit flies, leeches, zebrafish, and humans . These high levels of similarity—or homology—in gene sequences among markedly different organisms can only be due to common ancestry, or homology.

DNA currently is the focus of much research, but one must not overlook that its essential role is to determine protein amino acyl sequences and, therefore, their structure and function (cf. Figure 18). Proteins are the molecules that perform the functions programmed by the genome. Among other types, proteins are enzymes, ion transporters; proteins involved in signal transduction and gene expression, etc. In animals, proteins constitute a great proportion of the body mass as part of the muscles.

In this exercise (adapted from BioRad, 2001) you will approach the study of evolutionary relationships among several groups of fish through the analysis of the protein composition of their muscles.

Protein analysis will be accomplished by SDS-PAGE (polyacrylamide gel electrophoresis in the presence of the denaturing agent sodium dodecyl sulfate), a technique in which proteins are separated based on their size (molecular mass). You will compare a traditional phylogenetic tree of fish groups (Figure 15) with the phenetic relationships among the fish studied here as inferred from *your data*.

DOI: 10.1201/9788743813224-7

NOTES

Part A. Experimental Procedures

1. Sample preparation

- Following the directions of your lab instructor, setup electrophoresis rig (Mini-Protean 3 or equivalent) and add (1×) TGS (Tris-Glycine-SDS) electrophoresis buffer to the chamber.
- Your instructor will provide you with several fish muscle samples, 20 μg each. Check the labels and write the identity of the samples on Table 10.
- Samples have been prepared in a solution like the SDS-containing (1×) TGS electrophoresis buffer. SDS is a detergent that facilitates the separation of proteins by electrophoresis because, besides being a denaturing agent, it gives them a negative electrical charge, hence they all migrate towards the anode (+).

2. Electrophoretic separation of fish muscle proteins

- Request the prestained protein molecular weight (MW) standards from your instructor.
- Using a 20 μL pipettor, load your gel in the order shown in Table 13 (4 lanes/group, unless your instructor indicates otherwise).

Table 13

Lane	**Volume (μL)**	**Sample**
1	Empty	—
2	20	Protein MW standards
3	20	Fish muscle sample 1:
4	20	Fish muscle sample 2:
5	20	Fish muscle sample 3:
6	20	Fish muscle sample 4:
7	20	Fish muscle sample 5:
8	20	Fish muscle sample 6:
9	Empty	—
10	Empty	—

- Run the electrophoretic separation for ~1 h at 200 V in (1×) TGS buffer or as indicated by your instructor.

Staining

- After electrophoresis, remove the gel from the cassette and transfer the gel to a container with 40 mL of Coomassie Blue stain.
- Stain the gel for 20–30 min, with gentle shaking for best results.
- Discard the stain in the container located in the hood and destain the gels in a large volume of destaining solution for at least 30 min to overnight (in which case you will coordinate with the laboratory staff to return to the lab to change the destaining solution). The destaining solution may need to be changed during the process.
- It helps to add a few tightly scrunched Kimwipes to adsorb the Coomassie Blue stain. Just be careful to avoid the paper touching the gel.
- Protein standards and blue-stained protein bands will be visible on a clear background after destaining.

4. Imaging

- Document your gel using one of the following methods (as directed by your instructor):

  ✓ Take a picture using the lab's camera or a cellular phone.
  ✓ Scan the gel by placing a transparent support underneath (such as an acetate sheet or transparency) and another one on top (to protect the scanner).
  ✓ Put the gel back to the large Petri dish or as indicated by the instructor.

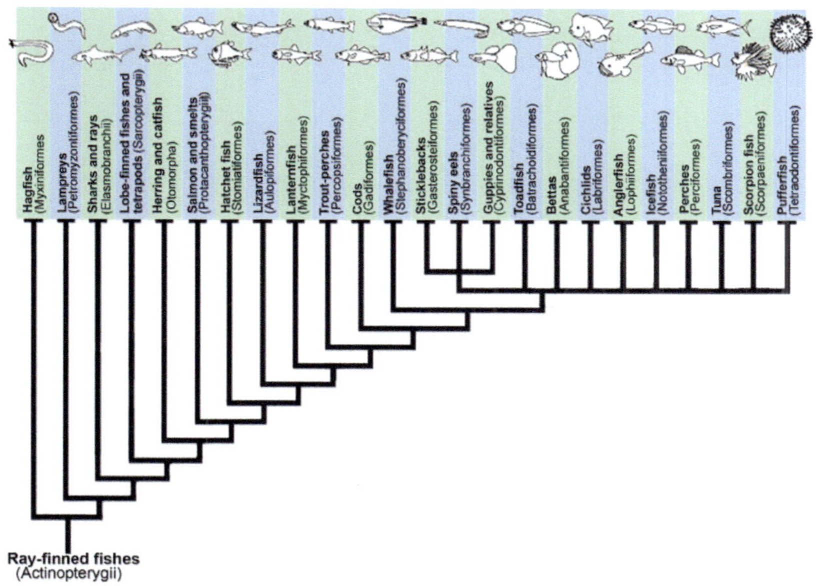

Figure 15. Phylogenetic tree of ray-finned fishes (Understanding Evolution, 2020).

NOTES

Part B. Data analysis

1. Phylogeny of fish groups.

- Familiarize yourself with the phylogeny and evolution of fish groups on the internet. You may visit the following web sites:

  Timetree: timetree.org. Find the function ***Build a Timetree*** and fill the window ***Specify a Group of Taxa*** with the term "Actinopterygii." Click the ***Search*** button.
  The Fish Tree of Life: fishtreeoflife.org. Click the link ***Browse our new "Phylogenetic Fish Classification"*** to explore the orders of bony (ray-finned) fishes.
- Study the phylogenetic tree in Figure 15 and observe the positions of the fish from which the protein samples were obtained.
- Notice how fish can be classified, according to morphological, physiological and other criteria, in relatively different and sometimes distant groups. For example, sharks are classified in a branch totally separate from other fish because their skeletons are cartilaginous (hence the name Chondrichthyes) and not bony as they are in the fish in the Ostheichthyes group. The Agnatha group includes the lampreys. Members of this group are characterized by the absence of a mandible, which is considered to be a primitive character (that is, a trait that precedes more modern ones). Notice how the Agnatha branch has been drawn on the diagram below the Chondrichtyes and also how both Agnatha and Chondrichtyes constitute phylogenetic groups as separate between themselves and other fish as they are from the other four main classes of Chordates.
- Hypothesize which fish species from which you obtained samples will have similar patterns of muscle proteins.

2. Comparison of protein composition.

- Use the picture or scan of your SDS-PAGE separation of fish muscle proteins and measure the distance from the application well to each one of the molecular weight (MW) marker bands. The small proteins will move further down the gel. Use Table 14 to identify the proteins and write down their migration distance.

Table 14

Protein MW Standards	Molecular weight (kDa)	Distance migrated from well (mm)
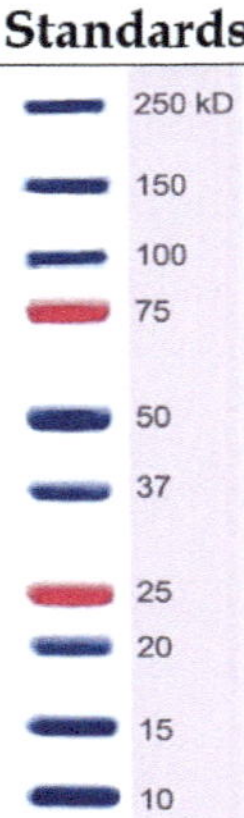 250 kD	250	
150	150	
100	100	
75	75	
50	50	
37	37	
25	25	
20	20	
15	15	
10	10	

- Construct a standard curve on semi-logarithmic paper with three cycles (see example in Figure 16) by plotting size (molecular weight) of standards in kilodaltons (kDa) in the log axis (Y) versus migrated distance (mm) in the linear axis (X).

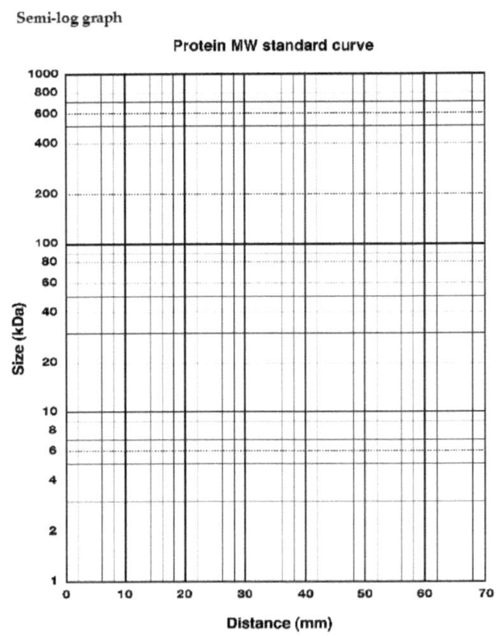

Figure 16

- In your gel, the actin/myosin-laden sarcomere appears as myosin (210 kDa) and actin (43 kDa). The major band below actin is tropomyosin, with a MW of 40 kDa and several myosin light chains are below 30 kDa.
- Observe that some bands are prominent across the gel, but their positions are slightly different. Their abundance may also vary, as evidenced by the intensity of their staining.
- Notice which samples are more similar than others and fine-tune your analysis by determining the molecular weight of those protein bands that vary from sample to sample and those that are common to two or more samples.
- Please note that it is difficult to assign identities to the proteins in your gel, since the fish muscle extracts contain complex mixtures of proteins. However, since the samples are all from muscle tissue, one may safely assume that there would be large quantities of the predominant proteins, i.e., actin and myosin. Definitive identification of a protein would require sequencing its amino acid residues, mass spectrometry, or immunodetection with specific antibodies.
- Molecular weights are estimated by measuring the distance migrated by the bands from the well (in mm) and interpolate this value in the X-axis (distance) of your standard curve; find the corresponding size (kDa) on the Y-axis (size).

Questionnaire
Lab 7. Molecular Evolution

Your report must include (a) a print-out of the stained polyacrylamide gel; (b) protein standard curve (this page); (c) a table containing estimates of protein size for each sample (see format below); and (d) answers to the questionnaire on the following pages.
Format of table for comparison of protein samples:

Fish Samples	**Molecular weight (kDa)**																		
Most similar																		…	n
Sample being compared																		…	n
Least similar																		…	n

Use the gel in Figure 17 as a guide to compare most and least similar protein bands:

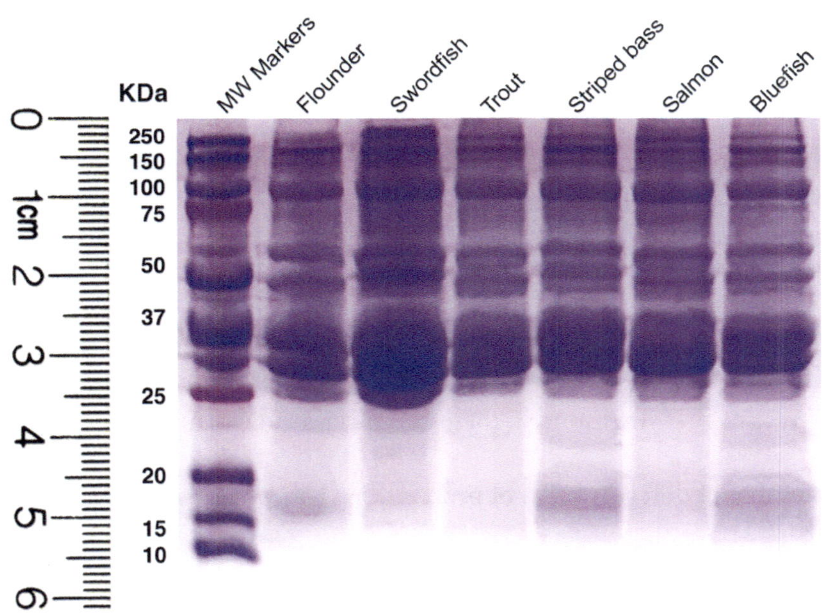

Figure 17. Polyacrylamide gel stained with Coomassie Blue, revealing the protein bands from fish muscle samples. Please notice that the standard ladder may be different from the one you are using.

Comparing the molecular weights of your proteins across the gel, confirm or reject the hypotheses that you elaborated while working with the phylogenetic tree regarding how close or how far the fishes you used as sample sources are from each other.

Discussion guide

1. List the names of the fish whose muscle proteins you are investigating.

2. According to the phylogenetic tree, which of your fish are most closely related to each other? Which fish are most distantly related to each other?

3. Which samples shared the most protein banding similarity? The least protein banding similarity?

4. How did you distinguish the protein profiles of different species from each other?

5. What are possible explanations for this variation?

6. Before you conducted the investigation, which two fish species did you list as being most related?

7. Of all the muscle proteins that you found in these two species, how many are present in both species?

8. What is the total number of different kinds of proteins that you were able to detect on your gel in these two species?

9. Of the total number of proteins in this pool, how many are found in common to both species listed in question 6 above?

$$\left(\frac{\text{N}^\circ \text{ of proteins in question 7}}{\text{N}^\circ \text{ of proteins in question 8}}\right) \times 100 = \_\_\_\_\%$$

10. Prior to starting this laboratory, which two fish species did you think were the least related?

11. Of all the muscle proteins that you found in these two species, how many are present in both species?

12. What is the total number of different kinds of proteins that you were able to detect on your gels, in these two species?

13. What percent of the muscle proteins were common to these least similar species?

$$\left(\frac{\text{N}^\circ \text{ of proteins in question 11}}{\text{N}^\circ \text{ of proteins in question 12}}\right) \times 100 = \_\_\_\_\%$$

14. Do your protein size data confirm your hypotheses about the relationships of the fishes you worked with? Do your data correlate with the arrangement of branches of the evolutionary tree?

15. What do the relative positions of the bands on the gel indicate about the proteins in the bands?

16. Are all the bands of equal thickness?

17. How would you explain the observation that some proteins form thin bands while others form thick bands?

18. Actin and myosin are proteins found in muscle tissue of all animals. Based on your data, what can you say about these proteins in the fishes you have investigated?
What might you find if you looked at actin and myosin in other animals?

19. Based on the apparent molecular weights of actin from your gel, predict the number of amino acids in this protein using an average molecular weight of 110 Daltons per amino acid.

20. What implications could the kind of molecular data like these protein comparisons have in relation to the theory of evolution? (In terms of support, rejection, neutrality, etc.)

The Genetic Code

Second letter

First letter	U	C	A	G	Third letter
U	UUU Phe	UCU Ser	UAU Tyr	UGU Cys	U
U	UUC Phe	UCC Ser	UAC Tyr	UGC Cys	C
U	UUA Leu	UCA Ser	UAA Stop	UGA Stop	A
U	UUG Leu	UCG Ser	UAG Stop	UGG Trp	G
C	CUU Leu	CCU Pro	CAU His	CGU Arg	U
C	CUC Leu	CCC Pro	CAC His	CGC Arg	C
C	CUA Leu	CCA Pro	CAA Gln	CGA Arg	A
C	CUG Leu	CCG Pro	CAG Gln	CGG Arg	G
A	AUU Ile	ACU Thr	AAU Asn	AGU Ser	U
A	AUC Ile	ACC Thr	AAC Asn	AGC Ser	C
A	AUA Ile	ACA Thr	AAA Lys	AGA Arg	A
A	AUG Met	ACG Thr	AAG Lys	AGG Arg	G
G	GUU Val	GCU Ala	GAU Asp	GGU Gly	U
G	GUC Val	GCC Ala	GAC Asp	GGC Gly	C
G	GUA Val	GCA Ala	GAA Glu	GGA Gly	A
G	GUG Val	GCG Ala	GAG Glu	GGG Gly	G

Figure 18. The genetic code is usually expressed as RNA codons, which occur in messenger RNA (mRNA). These codons are read during ribosomal protein synthesis, in a process known as translation.

NOTES

NOTES

NOTES

Laboratory N° 8
Genetic Fingerprinting

Introduction

The DNA sequences of human genomes are 99.9% identical among individuals. Although the DNA from different individuals is more alike than different, there are many regions of the human chromosomes that exhibit a great deal of diversity. Such variable sequences are termed polymorphic (meaning "many forms") and provide the basis for the diagnosis of some inherited diseases, forensic identification, and paternity testing. In this experiment, the now ubiquitous polymerase chain reaction (PCR) will be used to amplify a small DNA region from chromosome 8 to look for an insertion of a short nucleotide sequence called *Alu*. This *Alu* sequence may appear within the tissue plasminogen activator (TPA) gene.

Principle of the Polymerase Chain Reaction

PCR is used to make many copies (i.e. to clone or amplify) of specific fragments or regions of DNA (Figure 19). Two oligonucleotides (short single-stranded DNA fragments) are used as primers for a series of synthetic reactions that are catalyzed by DNA polymerase. These primers (one "forward" and one "reverse") are complementary to sequences found in the DNA fragments of interest that we want to amplify. The oligonucleotide primers possess two important characteristics: (1) they correspond to opposite strands of the DNA template and (2) they flank a segment of DNA that is to be amplified in the polymerase reaction.
The polymerase chain reaction contains a mixture of template DNA, oligonucleotide primers, DNA polymerase, and the four DNA nucleotide triphosphates (dNTPs) in a buffer solution. The amplification reaction consists of three separate stages:

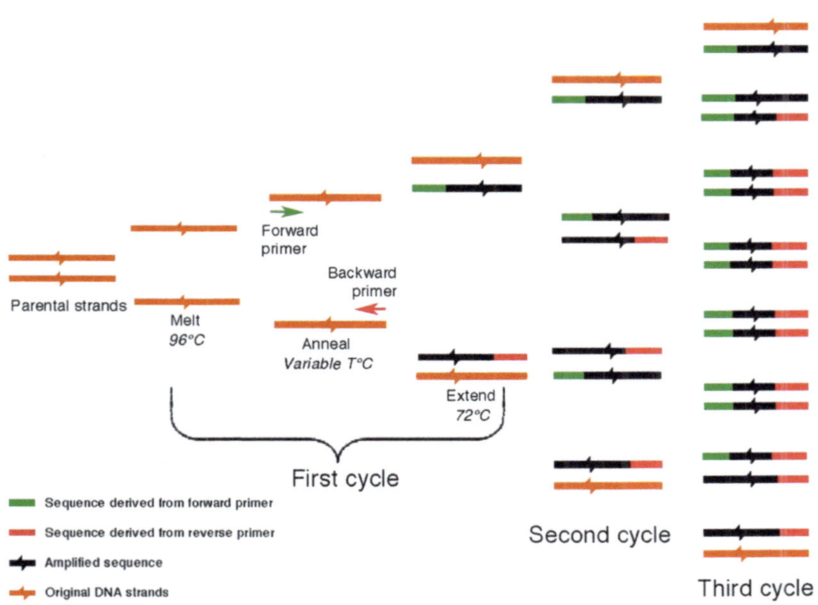

Figure 19. Schematic representation of the polymerase chain reaction.

DOI: 10.1201/9788743813224-8

First, the reaction mixture is heated at high temperature (94 °C) to separate the double-stranded DNA template into two complementary strands, a process known as denaturation or melting.
Second, the temperature is lowered (to a variable value, such as 58 °C) to allow the oligonucleotide primers to anneal with their corresponding sequences on the single-stranded DNA template. This annealing temperature is dependent on the nucleotide composition of both primers.
Third, the temperature is raised to the optimal temperature for the synthesis of the complementary DNA strand by the DNA polymerase (usually 72 °C). In this third step, the polymerase will use the annealed oligonucleotide primers to synthesize single stranded DNA molecules that are complementary to the template strands and precisely the size of the DNA between the 5' ends of the primers.
These steps of denaturation, annealing, and DNA synthesis are repeated many times with the same reaction mixture. Because the products of one round of amplification serve as templates for the next round, each successive cycle doubles the amount of the desired DNA product. Therefore, PCR results in an exponential amplification of a specific double-stranded DNA segment whose termini are defined by the two oligonucleotide primers. In practice, PCR can result in an amplification level of 10^6 after 25–30 cycles of the reaction, namely, the number of copies of DNA synthesized is given by $y = 2^x$, where x = number of reaction cycles (cf. Figure 20, based on an experiment in which the amplification of a 564 bp DNA fragment was through a variable number of cycles).
Notice that the temperatures used at each step in PCR are very high compared to the 37 °C optimum for most enzymes purified from mammalian or bacterial sources, or around 20 °C, for enzymes from fungal and plant origin. High temperatures would normally denature an ordinary DNA polymerase. Thermostable DNA polymerases purified from the bacterium *Thermus aquaticus* and other extremophiles, are utilized in PCR. These organisms live in hydrothermal vents, such as the ones that are present in Yellowstone National Park. The temperatures of the vent waters are typically 95 to 100 °C (boiling water) and the enzymes from these organisms exhibit unusual thermotolerance.
The DNA sequence we want to amplify is the *Alu* insertion in the TPA gene as per protocol by Cold Spring Harbor Laboratory (1994).
The *Alu* family of repeated DNA sequences is found throughout the genomes of primates. *Alu* elements are approximately 300 bp in length and their name derives from their bearing a single recognition site for the endonuclease *Alu* I located near the middle of this 300 bp sequence. Over the past 65 million years, the *Alu* sequence has been amplified in humans to about 500,000 copies, comprising an estimated 5% of our genome. Perhaps 500–2000 of these *Alu* insertions are only found in *Homo sapiens*. Some insertions are so recent (approximately 1–2 million years ago) that they are not fixed in human populations.
Batzer et al. (1990) described a specific insertion of an *Alu* element within the tissue plasminogen activator gene. This *Alu* element, called TPA-25, is dimorphic meaning that it is present in some individuals but not others. The insertion of *Alu* occurs within an intron (a region of the gene whose transcribed RNA is eliminated before translation); this way, the presence of the *Alu* sequence does not affect the production of a complete TPA protein.

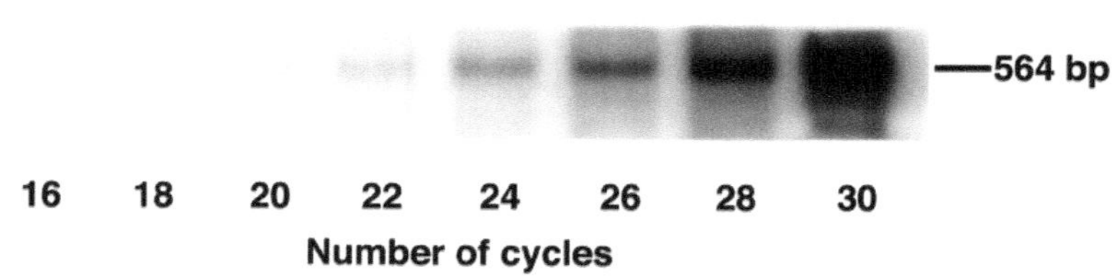

Figure 20. Effect of number of cycles on the amount of DNA amplified by PCR. The 564 bp fragment is an amplicon from the gene encoding VEGF (vascular endothelial growth factor) a cell-selective mitogen linked with new blood vessel formation (Suzuma et al., 2001).

Batzer's group used the polymerase chain reaction (PCR) to screen individuals for the presence of the TPA-25 insertion. Primers were designed to flank the insertion site and allow the amplification of a 400 bp fragment when TPA-25 is present. In the absence of TPA-25, the fragment is only 100 bp long.

There are three possible genotypes: homozygotes for the presence of TPA-25 (+/+), producing the 400 bp fragment only; homozygotes for the absence of TPA-25 (–/–), producing the 100 bp fragment only; and heterozygotes, producing both 400 bp and 100 bp fragments (+/–).

Individuals from each genotype would produce either a single band of 400 (+/+) or 100 nucleotides (–/–) in length, or two bands (+/–). Therefore, PCR products from homozygous and heterozygous individuals can be distinguished by amplifying the region of the TPA intron of interest followed by sizing in agarose gel electrophoresis.

Template DNA, from your two copies of chromosome 8, will be obtained by saline mouthwash, a painless, bloodless and noninvasive procedure. The cells are collected by centrifugation and resuspended in a solution containing the matrix Chelex, which binds metal ions that inhibit the PCR reaction. The cells are lysed by boiling and centrifuged to remove cell debris. A sample of the supernatant containing chromosomal DNA is mixed with *Taq* DNA polymerase, oligonucleotide primers, the four deoxyribonucleotide phosphates (dNTPs), and the cofactor Mg^{2+} (as $MgCl_2$).

Temperature cycling is used to denature the target DNA, anneal the primers, and extend a complementary DNA strand. As explained above, the size of the amplification products depends on the presence or absence of the *Alu* insertion at the TPA-25 locus on each copy of chromosome 8.

With the purpose of comparison of the genotypes from different individuals, aliquots of the amplified sample and those of fellow students are loaded into wells of an agarose gel. Following electrophoresis and staining, amplification products appear as distinct bands in the gel—the distance moved from the well is inversely proportional to the presence or absence of the TPA-25 insertion. Size is determined by comparison with standards run in the same gel. One or two bands are visible in each lane, indicating that an individual is either homozygous or heterozygous for the *Alu* insertion.

NOTES

Experimental Procedure

(A) Collection and Amplification of your DNA

Isolation of cheek cell DNA for PCR amplification

- Obtain a 50 mL conical culture tube containing 12 mL of 0.9% (w/v) NaCl solution. Pour all of the saline solution into your mouth and vigorously swish for 10 seconds; additional gently scraping with a sterile stick could be done as instructed. Do not discard the empty tube that contained the saline solution.
- Expel the sample solution back into the 50 mL sterile tube and close cap tightly.
- Spin the sample in a centrifuge at 1,100 × g for 10 min at 4 °C.
- Carefully, slowly pour off supernatant (liquid on top) into the sink and place the tube containing your cells (pellet) on ice.
- Using a 1000 µL micropipette, transfer 500 µL of Chelex suspension to the tube containing your cell pellet as follows:

  ⇒ Set the micro pipettor to 500 µL. Pipet up and down the Chelex solution several times to ensure the Chelex beads are fully resuspended.
  ⇒ Before the Chelex beads have time to settle, transfer 500 µL of the solution into the centrifuge tube containing the cell pellet.

- Mix the cells and Chelex by pipetting up and down several times until no visible cell clumps remain.
- Retain the pipette tip; do not discard it.
- Using the same pipette tip, transfer 500 µL of your resuspended sample into a clean 1.5 mL tube. Be sure to use a Sharpie to label the cap of the tube with your initials.
- Place your tube in a boiling water bath for 10 min.
- Carefully remove your tube from the boiling water bath and place on ice for 1 min.
- Place your tube in a microcentrifuge (opposite someone else's tube or a tube with the same volume of water) and centrifuge for 30 s. If the number of tubes is not even, place a "balance" tube containing 500 µL of Chelex bead suspension.
- Use a fresh pipette tip to transfer 200 µL of *supernatant* (the clear solution on top) to another clean 1.5 mL tube. Using a Sharpie, label with your name. Be careful not to transfer any Chelex or debris from the bottom pellet in the tube. Place the tube on ice.
- Mark the lid of a 0.5 mL microcentrifuge tube with your initials.

Amplification of *Alu* sequences by PCR

Your instructor will explain and demonstrate the following steps.

- Add the following components to a 0.5 mL microcentrifuge tube:

  - 2.5 µL cheek cell DNA preparation
  - 22.5 µL PCR reaction mixture (provided by instructor) which includes the oligonucleotide primers, *Taq* polymerase, dNTPs, buffer and the cofactor $MgCl_2$.

DOI: 10.1201/9788743813224-9

- Settings for PCR master mix and primer stocks are described below under "Preparation of PCR Mix" (cf. Table 15).

- Mix the components gently and start the thermocycler program. The amplification protocol will be as follows:

$$94^{\circ}\text{C for 1 min followed by 30 cycles of} \begin{cases} 94^{\circ}\text{C for 1 min} \\ 58^{\circ}\text{C for 2 min} \\ 72^{\circ}\text{C for 2 min} \end{cases}$$

NOTES

(B) Interpreting your DNA Results

(B.1) Determining your TPA-25 Genotype

Your amplified DNA will be subjected to gel electrophoresis. Oligonucleotide primers, flanking the insertion site, were selected to amplify a 400 bp fragment when TPA-25 is present (represented as the + allele) and a 100 bp fragment when it is absent (the – allele). Each of the three possible genotypes, namely homozygotes (+/+) for the presence of TPA-25 (400 bp fragment only), homozygotes (–/–) for absence of TPA-25 (100 bp fragment only), and heterozygotes (+/–) (both 400 bp and 100 bp fragments) can be distinguished following electrophoresis in agarose gels.

You will receive a melted agarose suspension in a flask. Follow the instructor's directions for pouring the gel into the electrophoresis apparatus. Take care to prevent forming bubbles as you pour.

Mix 18 μL of PCR sample with 2 μL of sample-loading buffer. Follow the instructor's instructions for loading the sample into the well of the gel. Load molecular weight standards (100 bp ladder) as indicated by your instructor (depending on the brand and concentration, 5 or 10 μL).

At the end of the electrophoretic run, observe the stained gel containing your sample under UV light and compare it to those from others. First, ascertain whether or not you can see a diffuse (fuzzy) band of "primer-dimers" that appears at the same position in each lane toward the bottom of the gel. Primer dimers are not amplified human DNA but are an artifact of the PCR reaction that results from the primers amplifying themselves.

⇒ Draw a picture illustrating the gel results and make sure to scan it or photograph it.

⇒ Excluding the primer-dimers band (if present), interpret the allele bands in each lane of the gel:

- ☐ No bands visible. This usually results from an error during sample preparation, such as losing the cell pellet or using a too-acidic Chelex solution.
- ☐ One visible band. Establish whether it corresponds to the 400 bp or 100 bp PCR product. Remember, the 400 bp band migrates slower than the 100 bp band and is therefore located closer to the sample wells. If only the 400 bp band is present, then that individual is homozygous for the TPA-25 *Alu* insertion (+/+). If only the 100 bp band is present, then that individual is homozygous for the absence of the TPA-25 *Alu* insertion (–/–).
- ☐ Two visible bands. This individual is heterozygous for the TPA-25 *Alu* insertion (+/–).
- ☐ Three or more bands visible. The one or two bright bands are likely the true alleles. Additional bands may occur when the primers bind nonspecifically to chromosomal loci other than TPA-25 and give rise to additional amplification products.

Please notice that some of the questions below are also contained in the report, so working diligently here will make working on your report much easier.

⇒ Determine the genotypes of the student samples in your class. How many homozygous for the presence of the TPA-25 *Alu* insertion (+/+), its absence (–/–) and heterozygous (+/–)?

(B.2) Population Genetics

If an allele benefits an organism carrying it, its frequency may increase in the population over time. For example, in rabbits, an allele that results in a camouflage fur color becomes more common than an alternative conspicuous color allele. This is because more of the camouflage-colored rabbits survive long enough to have offspring who will then inherit the allele and pass it on. The other allele decreases as those rabbits with the non-camouflage color are killed by predators before they pass along the allele.
Not every beneficial allele increases rapidly over time. In humans the Δ32 allele gives protection against HIV. However, this allele is not noticeably increasing in frequency. This may be because the allele is so rare in populations with high HIV rates. In fact, Δ32 is most common in Scandinavia and Scotland, where is found in about 10% of the population. These are places where this allele once protected those people against the Bubonic Plague in the 1300s! It's just a coincidence that the same protection works against HIV, but very few carriers of Δ32 are exposed to HIV.
The benefits or costs of having an allele can be studied by using population genetics. An allelic frequency is a ratio that compares the number of copies of a particular allele to the total number of alleles present in the population. If a population is genetically stable, the allelic frequencies will remain constant from one generation to the next. Such a population is said to be in Hardy-Weinberg equilibrium.

- Because humans are diploid, the total number of alleles present in your class equals the total number of students times two. Determine this number for your class.
- Your instructor will record the genotype for each group on the board and the number of students in each group. Assume that every member of each group has the same genotype. Record this information for your class population.

  For example, for a class of 20 students, their genotype distribution was as follows:

  - 5 students with (+/+) genotype (homozygous for the + allele)
  - 11 students (+/–) genotype (heterozygous)
  - 4 students = (–/–) genotype (homozygous for the – allele).

- Determine the total number of + alleles by counting the number of students who are homozygous for the allele and multiplying by 2, and counting the heterozygous students once, since they carry only one copy of the + allele.

  Number of + alleles in the example above: ________

  Number of + alleles in your class:________
- The allelic frequency is the ratio of the number of a particular allele to the total number of alleles in the population. Determine the allelic frequency for the + allele.

  For the example above: _________

  For your class: _________
- What is the allelic frequency for the – allele?

  For the example above: _________

  For your class: _________

Because your class population is so small it doesn't reveal much about how common the TPA-25 *Alu* insertion is and whether it is decreasing or increasing. However, this type of analysis is used to track changes in the frequency of more important alleles such as Δ32 or other alleles that might affect human populations over time.

(B.3) Practical Applications

Forensics

The allele we are examining occurs too frequently to be used in DNA fingerprinting. However, if the (–) TPA-25 *Alu* allele only occurred in one out of 1000 people, and if we also uncovered another genotype which occurs once in 1000 people, then those 2 alleles would only be expected to occur together once in $10^3 \times 10^3$, i.e. once in 1 million people (or a frequency of $1/10^6$ or 10^{-6}).
If they were found together in DNA collected at a crime scene, and if the suspect also had that rare combination of alleles, this might be used in a trial to "prove" guilt. However, the defense could argue that it was just a coincidence (as in highly publicized trials in recent years), because in a country with 300 million people, you would expect about 300 people to have that combination. So, the DNA at the crime scene might belong to any one of those 300 people.

(a) What is the practical problem with this argument? How convincing is DNA fingerprinting at proving guilt?

(b) On the other hand, if the suspect *does not* have the allele combination present at the crime scene, can anything be concluded about the suspect?

NOTES

Preparation of PCR mix

Primers

Forward (sense) primer
5'-GTAAGAGGTTCCGTAACAGGACAGCT-3'

Reverse (antisense) primer
5'-CCCCACCCTAGGAGAACTTCTCTTT-3'

Working solution: 12.5 μM of each primer. Preparation of stock primer solutions is done depending on the scale of primer available (i.e., 100 nmole, 1 μmole, 5 μmole, etc.).
Including 1 μL of each primer in a 25 μL volume reaction, renders a final concentration of each primer of 0.5 μM (cf. Yasin et al., 2019).

PCR master mix

Table 15

(volumes in μL)

Component	**Single tube**	**5 reactions**	**10 reactions**	**20 reactions**
Commercial (2×) PCR mix	12.5	62.5	125	250
Forward primer	1	5	10	20
Reverse primer	1	5	10	20
diWater	8	40	80	160
Master mix total volume	22.5	112.5	225	450

Final 22.5 μL volume of PCR master mix is added with 2.5 μL of cheek cell DNA prep per tube.

Note: Primer stocks may be prepared at a concentration of 10 μM each. Hence, 1 μL of each primer in a 25-μL reaction will render a final concentration of 0.4 μM each.

Questionnaire
Lab 8. Genetic Fingerprinting

Part 1. Determining the TPA-25 genotypes of students in your class

i. Paste a scan (or a photograph) of your gel. You may include only the gel region containing the MW markers (100 bp ladder) and the amplicon bands.

ii. Excluding the primer-dimers band (if present), interpret the "allele" bands in each lane of the gel and determine the genotypes of the student samples in your class. How many homozygous for the presence of the TPA-25 *Alu* insertion (+/+) [______] or its absence (–/–) [_______] and heterozygous (+/–) [_______]?

iii. Because humans are diploid, the total number of alleles present in your class equals the total number of students times two. Determine this number for your class.

TPA-25 "allele" present (+) = N° (+/+) + N° (+/–)/2 = ______

TPA-25 "allele" absent (–) = N° (–/–) + N° (+/–)/2 = ______

Total ______

iv. The allelic frequency is the *ratio* of the number of a particular allele to the *total* number of alleles in the population. Determine the allelic frequency for the + allele for your class.

v. Determine the allelic frequency for the – allele for your class.

vi. Assume that the TPA-25 *Alu* insertion is in Hardy–Weinberg (H-W) equilibrium. Use the W-H model to determine the allelic frequency of the alleles + and – in your class (you already did something similar in Lab 1. Bioinformatics).

Recall that the H-W equation is quadratic ($1 = \mathbf{p}^2 + 2\mathbf{pq} + q^2$); **p** and **q** are the frequencies of the alleles involved and that $\mathbf{p} + \mathbf{q} = 1$.

vii. Do your results for **p** and **q** coincide with your answers in the previous questions?

❑ Yes ❑ No

If they do not, what could be the reason(s)?

NOTES

Part 2. Forensic Applications

i. How convincing is DNA fingerprinting at proving guilt if used as forensic evidence? If you consult any bibliographic source, do not forget to cite it.

ii. If a suspect in a crime *does not* have the allele combination found in evidence collected at a crime scene, can anything be concluded about the suspect? Read again part (B.3) if necessary.

iii. The mouth of a bottle found at the scene of a crime has a sufficient number of epithelial cells from which enough nucleic acids can be extracted and analyzed by PCR. Shown in Figure 21 are the results of genotyping for three probes (loci 1, 2 and 3) of the evidence (X) and cells from five suspects (A–E). Notice that all samples were run on the same gel, but the technician posted a picture of the agarose gel on her notebook and properly labeled all lanes and probes used. Notice that the molecular weight markers (MW, in kb) are different from the ones you used in your PCR experiment.

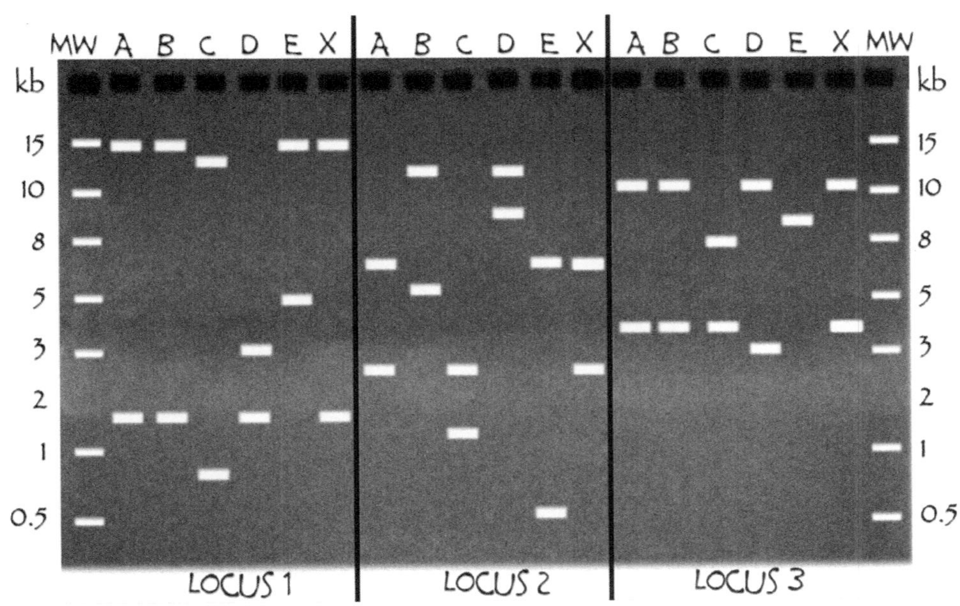

Figure 21. Schematic representation of the polymerase chain reaction.

Although this is only one piece of evidence, which suspect *cannot be excluded* as a perpetrator? ___
Explain.

NOTES

NOTES

Laboratory N° 9
Molecular Biology of Sickle-cell Anemia

Purpose

To separate, by electrophoresis in agarose gels two allelic forms of hemoglobin: the "normal" molecule, HbA, and the form found in people with sickle-cell anemia, HbS (Fox and Gaynor, 1996).

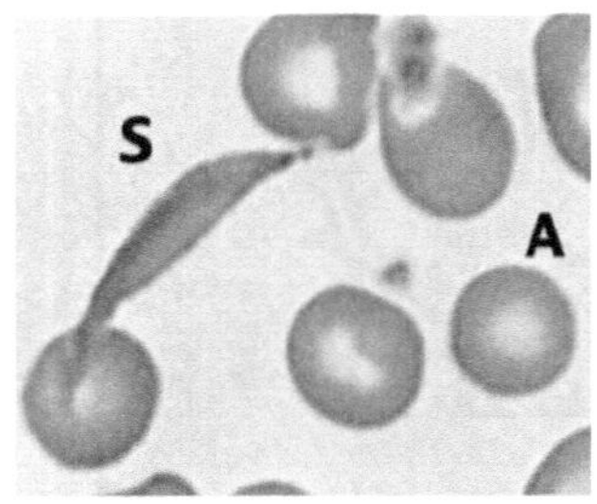

Figure 22. Erythrocytes containing (A) hemoglobin A, with non-mutated β chain, and (S) hemoglobin S. (Adapted from Encyclopædia Brittanica Online.)

Introduction

In previous laboratories we have used electrophoresis to separate protein mixtures (PAGE-SDS) or DNA samples (agarose gels). In this exercise will use agarose gels to separate two allelic forms of the red-blood-cell protein hemoglobin, responsible for oxygen exchange in higher animals.

In the adult human, hemoglobin is a tetramer composed by two each of two different types of polypeptides, α and β, hence Hb = $\alpha_2 + \beta_2$. The most common form ("normal") is hemoglobin A (HbA). The change of a single nucleotide in the gene for the β subunit causes the substitution of an amino acid residue: Glutamic acid, at position 6 of the β chain in HbA, has a negatively charged lateral group ($R = -(CH_2)_2-COO^-$).

A point-mutation in the hemoglobin gene in which the middle adenine in the codon for glutamic acid (GAA) is substituted by uracil (resulting in GUA), causes it to change to valine, which has a nonpolar lateral group ($R = -CH-(CH_3)_2$). This mutated form can be notated[14] as HbS: β6 Glu $\rightarrow$ Val, or simply HbS (S for "sickle").

The allelic form HbS is found in people with sickle-cell disease and sickle-cell trait. When an individual has two copies of the HbS (meaning, they have received mutated genes from both parents), all hemoglobin tetramers contain mutated β subunits and the shape of the entire hemoglobin molecule is altered. Consequently, erythrocytes become sickle-shaped (Figure 22), a structure that makes the cells fragile and causes them to break more easily when going through capillaries, causing *sickle-cell disease*, also known as falciform anemia.

If only one allele of the mutated gene is present (i.e., the person is heterozygote for the character), both HbS and HbA will be produced, with a predominance of HbA. This condition is called *sickle-cell trait*, and it does not produce occlusion of blood vessels. In contrast with sickle-cell disease, sickle-cell trait does not adversely affect the individual's life expectancy.

[14] This mutation can also be written as Eβ6V (cf. Lab N° 1. Bioinformatics)

DOI: 10.1201/9788743813224-10

Procedure

Gel preparation

Follow instructions in the solutions needed section. Notice that we will be using Tris-glycine buffer *without the denaturing agent SDS*, since we are working with proteins, in native conditions, in an agarose gel matrix.

Preparation of hemoglobin samples

While agarose gel solidifies, prepare the samples in small Eppendorf tubes as in Table 16.

Table 16

Simulated blood samples	10 mg/mL Hemoglobin A	10 mg/mL Hemoglobin S	Water	Loading buffer
Normal individual	10 µL	—	10 µL	5 µL
Sickle-cell trait	10 µL	10 µL	—	5 µL
Sickle-cell anemia	—	10 µL	10 µL	5 µL

Gel run

- Fill the electrophoresis rig with Tris-glycine buffer, pH 9.4, as indicated by the instructor.
- After the gel is completely set, carefully remove the comb and mount the gel in the electrophoresis tank. Add just enough buffer to cover the gel.
- Load the samples in the order indicated in Table 16. Use a pipettor and be careful not to break the agarose at the bottom of the wells. Make sure to change tips between loadings.
- Carefully, plug the electrophoresis box to the power supply. Run the electrophoresis at 80 V (constant voltage) for 60 min or until the fastest-running hemoglobin band reaches 3/4 of the gel length.
- Wearing gloves, take the gel out of the electrophoresis chamber. Since the hemoglobin possesses a red *heme* group, there is no need to stain the gel.
- Notice the protein bands pattern and measure the distance (in mm) that they have migrated from the origin (i.e., the well in the gel).
- If possible, take a picture using a white light box (at first try an f/32 aperture and a shutter speed of 1/125 s or use your cell phone).

NOTES

Equipment required

Minigel apparatus (includes combs, casting tray, and cover)
Power supply
Micropipettes and tips
Digital camera and a light-table or flatbed scanner.

Preparation of solutions

25 mM Tris-100 mM Glycine, pH 9.4
3 g Tris base
7.2 g glycine
Dissolve in 900 mL distilled water and adjust pH to 9.4 with 10 N NaOH.
Adjust volume to 1 L with distilled water.

1.5% (w/v) Agarose gel (Sambrook and Russell, 2001)

- Weigh 1.5 g of powdered agarose and add to 100 mL of Tris–glycine buffer, pH 9.4, contained in a flask suitable for boiling.
- Swirl to mix and microwave at high power for 3 min (or until the mix starts boiling). Lower the microwave oven power to 2 or 3 and let simmer until agarose is completely dissolved (by visual inspection). If necessary, add sterile deionized water to compensate for lost volume.
- Cool the solution to about 60 ¼C, pour in the gel cast and place an appropriate comb. Make sure that there is at least 1 mm of agarose between the bottom of the comb's teeth and the base of the gel. Let it solidify at room temperature.

10 mg/mL Hemoglobin A
5 mg Hemoglobin A (Sigma H0267)
Resuspend in 0.5 mL distilled water

10 mg/mL Hemoglobin S
5 mg Hemoglobin A (Sigma H0392)
Resuspend in 0.5 mL distilled water

Loading buffer
5 mL glycerol
5 mL distilled water
Dispense in ten 1 mL aliquots and keep frozen at −20 °C until use.

Questionnaire
Lab 9. Molecular Biology of Sickle-cell Anemia

1. Compare the picture of your gel with the expected migration as per the actual picture of a gel on the left of Figure 23.

Your picture (or scan):

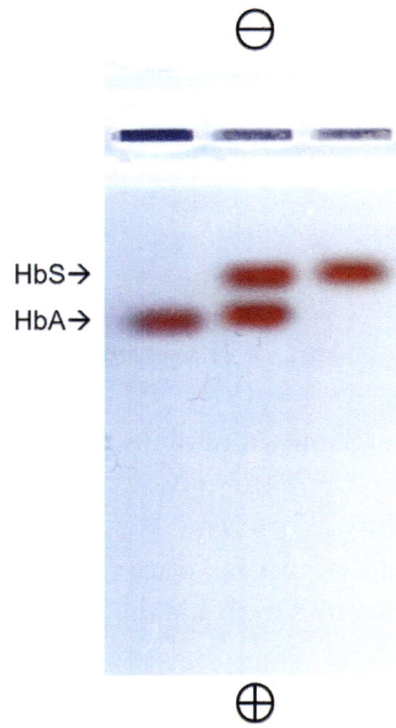

Figure 23

2. Consult a biochemistry textbook or handbook: What is the molecular weight of hemoglobin A *and* S (in kilodaltons, kDa)?
3. Does the change Glu→Val (Eβ6V) explain the unequal migration of HbA and HbS in your gel?
4. Enter the accession numbers for the β chain in hemoglobin alleles HB B (NM_000518.4) and HB S Wake (AY136510) in the nucleotide database at the NCBI website (www.ncbi.nlm.nih.gov/).

   Can you spot the difference(s) in the nucleic acid sequences for these alleles?

   ❑ Yes ❑ No

   If not, try a comparison between these two sequences at blast.ncbi.nlm.nih.gov/

NOTES

5. To learn more about hemoglobin-related disorders, check the entry for information about the β-hemoglobin's gene at www.ncbi.nlm.nih.gov/gene/3043.

   What other diseases are associated with extraneous forms of hemoglobin subunits in the human adult?
6. How do we call the type of point mutation in which an A→U change occurs in the codon for the sixth amino acid in hemoglobin chain β?

   ❑ Inversion ❑ Transversion ❑ Transition

7. What does the term "heterozygous advantage" mean?
8. What is the heterozygous advantage of people having sickle-cell *trait* in areas where malaria is a major cause of death?

NOTES

NOTES

Laboratory N° 10
Signal Transduction: Antagonistic Effects of Gibberellin and Abscisic Acid on the Production of α-Amylase in Barley Seeds

Purpose

To study the effects of the phytohormone gibberellic acid on the synthesis and release of α-amylase during the germination of cereal seeds and the inhibitory effects of abscisic acid (protocol adapted from Bush, 1994).

Introduction

The cereal seed is a good model for studies of the effects of hormones on germination. In cereals, the embryo is located on one side of the caryopsis, which makes it easy to excise. Other components of the seed are the starchy endosperm and the aleurone, a layer of cells that contains storage proteins. During germination (Figure 24), gibberellins (GAs) are produced in the embryo and transported to the aleurone cells (Firn and Kende, 1974). Once there, they trigger signal transduction pathways which eventually break dormancy, and lead to gene expression of enzymes required for mobilization of food reserves, for example, hydrolases involved in the degradation of starch in the endosperm and proteins in the aleurone itself.

Abscisic acid (ABA) has an important role in germination too, partly because seed dormancy is controlled by the ratio ABA/GA more so than the absolute concentrations of these phytohormones. ABA is a natural constraint that keeps developing embryos in their embryogenic state, but it also participates in other embryogenic processes; for example, it induces the synthesis of proteins that may have a role during periods of stress (Zhao et al, 2022). In contrast, during germination, ABA inhibits the GA-dependent synthesis of hydrolytic enzymes, including α-amylase, by suppressing the transcription of its mRNA.

Since the embryo is the source of both GA and ABA, detaching it from the rest of the seed allow us to add hormones to these embryoless seeds (which we will call "half-seeds"). In this way, the effects of these hormones could be distinguished from each other (Firn and Kende, 1974).

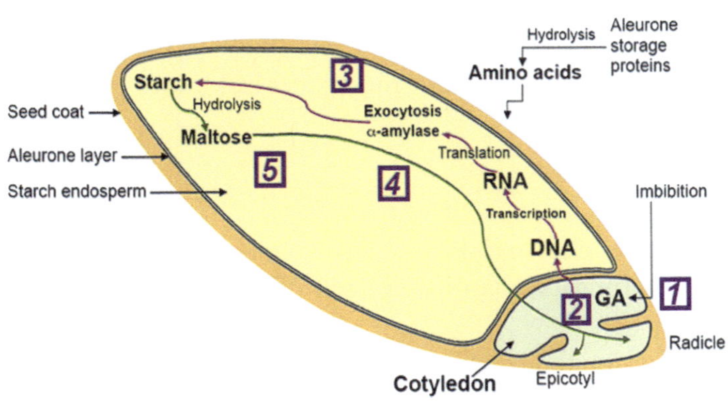

Figure 24. GA-induced mobilization of starch and other reserves in the cereal aleurone (Orth et al., 2025).

DOI: 10.1201/9788743813224-11

During germination of the barley seed, five events (Taiz and Zeiger, 2014) can be defined (Figure 24):

1. The embryo produces and releases GA.
2. GA migrates to the aleurone layer.
3. GA induces the biosynthesis of hydrolytic enzymes.
4. Hydrolytic enzymes are secreted from the aleurone layer into the endosperm.
5. Starch is hydrolyzed to simple sugars.

In this laboratory, we are going to study the GA-induced production of α-amylases (EC 3.2.1.1) in the aleurone of barley (*Hordeum vulgare*), a layer of two to four tiers of cells located below the test (pericarp) layers of the seed. α-Amylase degrades amylose, a component of the starch in the endosperm, to maltose, a disaccharide that is more readily mobilized into the embryo cells to serve as an energy source. The effect of adding ABA in conjunction with GA will also be studied. Specifically, we will vary the amounts of gibberellic acid and corroborate the hypothesis that the production of α-amylase is a function of the gibberellic acid concentration in the medium. Jones and Varner (1967) showed that the GA-induced release of α-amylase is proportional to the logarithm of the concentration of GA applied to barley half-seeds. By interpolation, one may be able to determine the amount of GA present in unknown samples, plant-derived or otherwise.

Procedure

(Please note, Parts 1-3 are usually prepared by the instructor or a technician)

Part 1. Preparation and imbibition of half-seeds

Date: ________ Time: ________

- Obtain 175 grains of huskless barley[15] (*Hordeum vulgare* cv. Himalaya; _____ harvest).
- Using a razor blade and a cutting board, cut them transversely, eliminating only the portion carrying the embryo.

  From this point on, work in a laminar flow hood or under a flame to preserve sterility
- Place the half-seeds in a 100 mL sterile beaker containing 50 mL of a 10% (v/v) solution of commercial bleach. Stir with a magnetic bar for 20 min. Pour off the bleach solution (be careful not to spill it on your clothes!) and quickly add 50 mL (or more) of sterile diH_2O. Stir and let it sit for about 5 min.
- Pour off water and wash at least twice more as described. Be careful not to throw the half-seeds away with the water.
- Transfer surface-sterilized seeds onto one layer of sterile filter paper inside a 10 cm Petri dish moistened with 5 mL of sterile diH_2O.

  Notes:

  - Do not place more than 50 seeds per plate and do not exceed the water volume.
  - Work with great care: bacterial or fungal contamination will affect and invalidate your results, since some microorganisms produce GA.
  - Cover the Petri dishes with aluminum foil and incubate at room temperature for 3 to 4 days to allow imbibition. Clean the sterile hood after you have finished your work.

[15] If barley seeds are not available, wheat can be used, although seed batches must be tested for sensitivity to GA.

NOTES

Part 2. Induction of α-amylase

Date: ________ Time: ________

- Table 17 indicates the final concentrations of hormones that are going to be used in the different treatments of barley half-seeds and if the experiment will include calcium (required for α-amylase activity) or not (Bush et al, 1989). Please analyze it.

Table 17

Flask N°	$CaCl_2$	GA_3	ABA	Unknown
1	—	—	—	—
2	5 mM	—	—	—
3	5 mM	—	50 µM	—
4	—	5 µM	—	—
5	5 mM	0.05 µM	—	—
6	5 mM	0.1 µM	—	—
7	5 mM	0.5 µM	—	—
8	5 mM	1 µM	—	—
9	5 mM	5 µM	—	—
10	5 mM	0.05 µM	50 µM	—
11	5 mM	0.1 µM	50 µM	—
12	5 mM	0.5 µM	50 µM	—
13	5 mM	1 µM	50 µM	—
14	5 mM	5 µM	50 µM	—
15	5 mM	?	—	?

- Work in sterile conditions. In the sterile Erlenmeyer flasks provided, add the following solutions in the amounts indicated in Table 18.

Table 18

All volumes in mL

Flask N°	diH2O	0.1 M $CaCl_2$	1 µM GA_3	10 µM GA_3	1 mM ABA	Unknown sample
1	2.0	—	—	—	—	—
2	1.9	0.1	—	—	—	—
3	1.8	0.1	—	—	0.1	—
4	1.0	—	—	1.0	—	—
5	1.8	0.1	0.1	—	—	—
6	1.7	0.1	0.2	—	—	—
7	0.9	0.1	1.0		—	—
8	1.7	0.1	—	0.2	—	—
9	0.9	0.1	—	1.0	—	—
10	1.7	0.1	0.1	—	0.1	—
11	1.6	0.1	0.2	—	0.1	—
12	0.8	0.1	1.0		0.1	—
13	1.6	0.1	—	0.2	0.1	—
14	0.8	0.1	—	1.0	0.1	—
15	1.8	0.1	—	—	—	0.1

- Transfer 10 sterile half-seeds under sterile conditions to each one of the flasks and incubate for 48 h at room temperature with continuous shaking at 40–60 cycles/min.

  Start: ______ Stop: ______

NOTES

Part 3. Preparation of incubation medium for measurement of α-amylase activity

Date:

- Measure volume of incubation medium and transfer to clean centrifuge tubes or 100 × 7.5 mm test tubes.
- Rinse flasks and seeds with some sterile diH_2O to make a total volume equal to 4 mL in each tube. (Be careful not to exceed volume.)

Part 4. Determination of α-amylase activity

(This is the part that students will perform in the laboratory)

Assay principle

We are going to measure the enzymatic activity of α-amylase by following the disappearance of its substrate, amylose. In the presence of potassium iodide, amylose forms a blue–purple complex with iodine. The higher the activity of α-amylase, the smaller the amount of amylose–iodine complex in the reaction tube. The intensity of the blue–purple complex can be determined spectrophotometrically by changes in absorbance at 625 nm (A_{625}). Decreases in A_{625} are proportional to the quantity of α-amylase present in the reaction mixture.

Performing the assay

(4.1) **Reagents blank**

- Make sure starch substrate is at room temperature (it should have been stored at 4 ∘C).
- Set apart five test tubes. Proceed, for each tube, as indicated in the leftmost column of Table 19.
- Start with "Trial number 1," using 3 mL diH_2O.
- If $A_{625} \neq 0.7$, repeat with the volume indicated for "Trial number 2."
- Increase the volume of water until you get $A_{625} \approx 0.7$.

Table 19

	Trial number							
	1	**2**	**3**	**4**	**5**	**6**	**7**	**8**
1. Place 1 mL starch in a test tube	√	√	√	√	√	√	√	√
2. Add 1 mL IKI solution	√	√	√	√	√	√	√	√
3. Add diH_2O	1 mL	2 mL	3 mL	4 mL	5 mL	6 mL	7 mL	8 mL[16]
4. Read A_{625} in spectrophotometer								

- Notice volume of water added to get $A_{625} \approx 0.7$:_____ mL.
- Measure room temperature: _______ ∘C.

[16]The maximum volume depends on the test tubes used and the concentration of starch substrate. If starch is too concentrated the volume can be reduced to 0.5 mL; the amount of lugol should be 0.5 mL.

(4.2) **Assay of sample with the highest activity (Table 20)**

- Assay first the sample presumed to have the highest activity, namely the medium coming from the flask with the highest GA concentration and no ABA.
- Try several sample volumes and/or incubation times. Start with sample volumes of 20 µL and short times (e.g. 30 s) and modify as necessary. For example, using 20 µL aliquots, increase the time; or maintain the time constant but use 40 µL of sample, then 60 µL, etc. The idea is to obtain a value of $A_{625} \approx 0.3$ for the sample with the highest activity.

Table 20

	Trial number 1	2	3	4	5
1. Place 1 mL starch in a test tube	√	√	√	√	√
2. Add sample (10–100 µL). *Write down vol. used*	µL	µL	µL	µL	µL
3. Incubate at room temperature (20 s–1 min). Time:	s	s	s	s	s
4. Add 1 mL IKI solution	√	√	√	√	√
5. Add diH_2O (vol. determined in step 4.1)	mL	mL	mL	mL	mL
6. Read A_{625}					

- Write down sample volume used: _____ µL, assay time: _____ min; and diH_2O volume: ______ mL.

(4.3) **Assay of samples**

- Using the sample volume and the assay time determined in (4.2), as well as the water volume determined in (4.1), proceed to assay *all* samples, including a new reagents blank.
- Take in account that, if your assay times are very short (say, less than one minute) you must work on one tube at a time. But remember, everything must be constant, except the source of your sample.
- Prepare your samples and write down your data in Table 21.

Table 21

α-Amylase activity determination						
Flask	**Starch substrate**	**Sample volume**	**Incubation time**	**IKI reagent**	**diH$_2$O**	**A_{625}**
Blank	1 mL	—	min	1 mL	mL	
1	1 mL	µL	min	1 mL	mL	
2	1 mL	µL	min	1 mL	mL	
3	1 mL	µL	min	1 mL	mL	
4	1 mL	µL	min	1 mL	mL	
5	1 mL	µL	min	1 mL	mL	
6	1 mL	µL	min	1 mL	mL	
7	1 mL	µL	min	1 mL	mL	
8	1 mL	µL	min	1 mL	mL	
9	1 mL	µL	min	1 mL	mL	
10	1 mL	µL	min	1 mL	mL	
11	1 mL	µL	min	1 mL	mL	
12	1 mL	µL	min	1 mL	mL	
13	1 mL	µL	min	1 mL	mL	
14	1 mL	µL	min	1 mL	mL	
15	1 mL	µL	min	1 mL	mL	

(4.4) **Determination of α-amylase units**

Jones and Varner (1967) proposed the following equation to calculate units of α-amylase:

$$UA = \frac{\Delta A_{625} \times T_v}{t \times v}$$

where:

ΔA_{625} = A_{625} water blank – A_{625} sample
Tv = volume of supernatant (mL) → Use 1 mL
t = time of incubation with starch (min) → Convert s to min
v = volume of supernatant tested (mL) → Convert µL to mL

Presentation of results

1. Calculate your α-amylase using a calculator or an Excel spreadsheet. Using Excel, construct a bar graph of UA versus treatments (this is **graph GA-1**).
2. Make a separate, semilog graph of UA from flasks 5 to 9 (UA in the ordinate vs. log [GA] in the abscissa, namely: 0.05, 0.1, 0.5, 1, and 5 µg/mL). This is **graph GA-2**.

   Note: Do not label the X-axis as "5, 6, ..., 9." Instead, indicate GA concentrations.
3. In another semilog graph, repeat the trace of graph GA-2 but add the data from flasks 10 to 14. Carefully indicate the GA concentrations on the X-axis and the legend "[GA] + 50 µM ABA." Remember that the X-axis should be logarithmic. This is **graph GA+ABA**.

   Note: Do not label the X-axis as "10, 11, ..., 14." Instead, indicate GA concentrations.
4. Flask number 15 contains an "unknown sample," which contains some GA_3. The idea is to estimate the concentration of GA_3 in this sample by comparison with the α-amylase activities in the "known samples."

   Interpolate the UA value from flask 15 in graph GA-2 and estimate the concentration of your unknown sample. Another way to do this is by determining the equation of your curve using a spreadsheet program and substituting the value of UA in it. This system is used as a bioassay for GA activity in extracts of plant materials or for synthetic gibberellins.

NOTES

Preparation of reagents

100 mM $CaCl_2$ stock
Dissolve 1.47 g of $CaCl_2 \cdot 2H_2O$ (f.w. 147) in diH_2O to a final volume of 100 mL.

5 mM GA_3 stock
Dissolve 19.2 mg of gibberellic acid, potassium salt (GA_3; f.w. 384.5) in 10 mL of 95% (v/v) ethanol.

10 µM GA_3
Dissolve 20 µL of 5 mM GA_3 stock in 10 mL diH2O.

1 µM GA_3
1 mL of 10 µM GA_3
Q.s. ad 10 mL with diH_2O

10 mM ABA stock
Dissolve 26.4 mg abscisic acid (f.w. 264.3) in 10 mL 95% (v/v) ethanol.

1 mM ABA
1 mL of 10 mM ABA
Q.s. ad 10 mL with diH_2O

0.05 N HCl
2 mL HCl (conc., 37%, f.w. 36.46, ρ = 1.19 g/mL)
Q.s. ad 500 mL with diH_2O

Iodine stock
Dissolve 6 g KI and 600 mg I_2 in diH_2O to a final volume of 100 mL.

IKI solution
1 mL Iodine stock
99 mL 0.05 N HCl

50 mM MES buffer stock, pH 5.6/20 mM Ca^{2+} (1 L)
Dissolve 10.7 g of 2-(N-morpholino)ethanesulfonic acid (monohydrate, f.w. 213.26) in about 700 mL of diH_2O.
Add 200 mL of 0.1 M $CaCl_2$ stock and adjust pH to 5.6 with 1N NaOH.
Q.s. ad 1 L with diH_2O.

Preparation of starch substrate

Dissolve 1.5 g of insoluble potato starch (*e.g.* Sigma S-4251) in 1 L of 50 mM MES buffer, pH 5.6/20 mM Ca^{2+}

↓

Boil for exactly 1 min

↓

Centrifuge at 4 °C for 10 min at 10,000 rpm

↓

Pellet
Discard

Supernatant
Decant onto a fresh container and keep refrigerated

↓

Let warm up to room temperature before use and return to refrigerator after use

Questionnaire
Lab 10. Signal Transduction

Discussion Guide

1. What is the purpose of working with embryoless seeds?
2. Was there any difference between A_{625} for the reagents blank versus that of flask **1**, the control with no additions, i.e., where half-seeds were incubated only with water?

   ❑ Yes ❑ No

   Explain.
3. Was there any difference in UA values for flasks **2** (no GA_3) and **3** (added with ABA)?

   ❑ Yes ❑ No

   Explain.
4. Flask **4** has 5 μM GA, but no Ca^{2+}. How does its α-amylase activity compare with that of flasks **9** (same [GA] + 5 mM $CaCl_2$) and **14** (same as flask 9 plus 50 μM ABA)?
5. Why is there α-amylase activity if Ca^{2+} was not added to the incubation medium?
6. From your graph GA-1, how is the production of α-amylase related to GA concentration (directly proportional; geometric proportion, no correlation)?

NOTES

7. Is that what you expected?

❑ Yes ❑ No

Why?

8. From your graph GA-2, what is the concentration of GA_3 in your unknown sample?

_____ AU → $[GA_3]$ = _____

9. From your graph GA+ABA, what was the effect of ABA on α-amylase activity? Is that what you expected? Why? Was it equally effective at all concentrations of GA?

NOTES

NOTES

Bibliography

Alaie A, Teller V, Qiu W-g (2012) A bioinformatics module for use in an introductory biology laboratory. Am Biol Teach **74**:318-332.

Batzer MA, Kilroy GE, Richard PE, Shaikh TH, Desselle TD, Hoppens CL, Deininger PL (1990) Structure and variability of recently inserted *Alu* family members. Nucleic Acids Res **18**:6793. Corrected in: Nucleic Acids Res **18**:699 (1991).

Bio-Rad (2001) "Biotechnology Explorer Protein Fingerprinting Instruction Manual," Prod. N¡ 166-0100EDU; BioRad, Hercules, CA.

Brooker RJ (2020) Genetics: Analysis and Principles, 7th Edition. McGraw-Hill. Pp. 106-130.

Bush, DS, Stich L, Van Huystee R, Wagner D, Jones RL (1989) The calcium requirement for stability and enzymatic activity of two isoforms of barley aleurone a-amylase. J. Biol. Chem. **264**:19392-19398.

Bush DS (1994) Plant Physiology Laboratory Manual. Department of Biological Sciences; Rutgers, The State University of New Jersey; Campus at Newark.

Cold Spring Harbor Laboratory (1994) "Detection of *Alu* by PCR: A Human DNA Fingerprinting Lab Protocol," as posted in Access Excellence: www.accessexcellence.org/AE/AEPC/DNA/detection.php, The National Health Museum, accessed March 13, 2008.

Firn RD, Kende H (1974) Some effects of applied gibberellic acid on the synthesis and degradation of lipids in isolated barley aleurone layers. Plant Physiol. **54**:911-915.

Fox M, Gaynor JJ (1996) A demonstration of the molecular basis of sickle-cell anemia. J Coll Sci Teach **26**:147-149.

Griffiths AJF, Wessler SR, Lewontin RC, Carroll SB (2008) Introduction to Genetic analysis, 9th Edition, WH Freeman, New York, p. 329.

Honts JE (2003) Evolving strategies for the incorporation of bioinformatics within the undergraduate cell biology curriculum. CBE Life Sci Educ **2**:233-247.

Horton HR, Moran LA, Ochs RS, Rawn JD, Scrimgeour KG (2002) Principles of Biochemistry, 3rd edition, Prentice-Hall, Upper Saddle River, NJ, p. 41.

Jones RL, Varner JE (1967) The bioassay of gibberellins. Planta **72**:155-161.

Leaver CJ, Hack E, Forde BG (1983) Protein synthesis by isolated plant mitochondria. Meth Enzymol 97:476-484.

Maloney M, Parker J, LeBlanc M, Woodard CT, Glackin M, Hanrahan M (2010) Bioinformatics and the undergraduate curriculum. CBE Life Sci Educ **9**:172-174.

Morral N, Bertranpetie J, Estivill X, Nunes V, Casals T, Gimenez J, Reis A, Varon-Mateeva R, Macek Jr. M, Kalaydjieva L, Angelicheva D, Dancheva R, Romeo G, Russo MP, Garnerone S, Restagno G, Ferrari M, Magnani C, Claustres M, Desgeorges M, Schwartz M, Schwartz M, Dallapiccola B, Novelli G, Fered C, de Arce M, Nemeti M, Kere J, Anvret M, Dahl N, Kadasi L (1994) The origin of the major cystic fibrosis mutation (ΔF508) in European populations. Nat Genet **7**:169-175.

National Center for Biotechnology Information (2005) NCBI Help Manual. URL: www.ncbi.nlm.nih.gov/books/NBK3831/Accessed: 11Jan25

Nelson DL, Cox MM (2021) Lehninger Principles of Biochemistry, 8th Edition. W.H. Freeman and Company, New York. Pp. 183-189.

New England BioLabs, Technical Reference, General Molecular Biology Data (www.neb.com/nebecomm/tech_reference/general_data/amino_acid_structures.asp) Downloaded: September 3, 2014.

Olczack M, Morawiecka B, Watorek W (2003) Plant purple acid phosphatases–Genes, structures and biological function. Acta Biochim Pol **50**:1245-1256.

Orth SH, Fox GP, Manley M, Hayward S (2025) Current detection methods and future trends for barley pre-harvest sprouting, is NIR the preferred method of the future? J Am Soc Brew Chem **83**:261-272

Qin L, Sharpe MA, Garavito RM, Ferguson-Miller S (2007) Conserved lipid-binding sites in membrane proteins: A focus on cytochrome *c* oxidase. Curr Opin Struct Biol 17:444-50

Sambrook J, Russell DW (2001) Molecular Cloning: A Laboratory Manual, 3rd edn. Cold Spring Harbor Laboratory Press, Cold Spring Harbor, New York. Pp. 5.4-5.13.

Schenk G, Gahan LR, Carrington LE, Mitic N, Valizadeh M, Hamilton SE, de Jersey J, Guddat LW (2005) Phosphate forms an unusual tripodal complex with the Fe-Mn center of sweet potato purple acid phosphatase. Proc Natl Acad Sci USA 102:273-278.

Siegel IH (1975) Enzyme Kinetics: Behavior and Analysis of Rapid Equilibrium and Steady-State Enzyme Systems. Wiley Interscience (Wiley Classic Library Edition, 1993), New York.

Siegel IH (1976) Biochemical Calculations, 2nd Edition, John Wiley & Sons, pp. 83-86.

Sigma Technical Bulletin CYTOCOX1. 2006. St. Louis.

Suzuma I, Hata Y, Clermont A, Pokras F, Rook SL, Suzuma K, Feener EP, Aiello LP (2001) Cyclic stretch and hypertension induce retinal expression of vascular endothelial growth factor and vascular endothelial growth factor receptor-2. Diabetes 50:444-454

Taiz L., Zeiger E, M¿ller I.A., Murphy A (2014) Plant Physiology and Development, 6th edn., Oxford University Press.

Tsui L-C (1992) Mutations and sequence variations detected in the cystic fibrosis transmembrane conductance regulator (CFTR) gene: A report from the cystic fibrosis genetic analysis consortium. Hum Mutat **1**:197-203.

Understanding Evolution: A Fisheye view of the tree of life. https://evolution.berkeley.edu/evolibrary/article/fishtree_01. Slightly modified for readability. Downloaded: June 17, 2020

Voet D, Voet JG (2004) Biochemistry, 3rd Edition. John Wiley & Sons, New York. Pp. 818-820

Wikimedia Commons https://commons.wikimedia.org/wiki/User:Sobarwiki Downloaded: 12Jan 25

Yasin SR, AlHawari HH, Alassaf AA, Khadra MM, Al-Mazaydeh ZA, Al-Emerieen AF, Tahtamouni LH (2019) Alu DNA polymorphism of human tissue plasminogen activator (tPA) gene in diabetic Jordanian patients. Iran Biomed J. **23**:423-428.

Zhao HY, Zhang Y, Zheng Y (2022) Integration of ABA, GA, and light signaling in seed germination through the regulation of ABI5. Front Plant Sci. 13:1000803.

Glossary

Amino Acid: fundamental building block of proteins

Amphipathic: a molecule having both hydrophilic and hydrophobic

Amphoteric: a compound able to react both as a base and as an acid

Anhydrous: a substance, especially a crystalline compound, containing no water

Anode: the positively charged electrode of a device by which electrons leave the device

ATP: Adenosine Triphosphate, primary energy-carry molecule in all living cells.

ATP synthetase: the enzyme that catalyze the synthesis of ATP

BLAST: Basic Local Alignment Search Tool that finds regions of local similarity between sequences

BSA: Bovine Serum Albumin

Coomassie Blue G-250: a dye that binds to hydrophobic regions of proteins

COX/Complex IV: Cytochrome *c* Oxidase

Cytoskeleton: internal skeleton of a cell that provide structural support through a dynamic network of proteins (actin) filaments and (micro-)tubules.

Database: an organized collection of structured data, stored electronically, designed for efficient access, management, and updating.

dH_2O: distilled water

diH_2O: deionized water

dihydrate: a chemical compound that incorporates two molecules of water into its crystal structure for every molecule of the compound

ddH_2O: double-distilled water

DNA: Deoxyribonucleic Acid

dNTP: deoxyribose DNA Nucleotide Triphosphate

ES: Enzyme-Substrate (complex)

$FADH_2$: Flavin adenine dinucleotide reduced

FASTA: a plain text format used in bioinformatics to store biological sequences (DNA, RNA, or amino acid) using single-letter codes

Hydrolysis: a chemical reaction in which a molecule of water is used to break one or more chemical bonds in a compound

Hydrophobic: tending to repel or fail to mix with water

Hydrophilic: having a tendency to mix with, dissolve in, or be wetted by water

kDa: Kilodaltons

K_m: Michaelis constant

Microtubules: tubular structures present in the cytoplasm of cells, can aggregate to form more complex structures that are part of the cytoskeleton

Monohydrate: a hydrate containing one mole of water per mole of the compound

Multiprotic: an acid that can donate more than one proton per molecule in acid-base reaction

MW: Molecular Weight

NADH: Nicotinamide Adenine Dinucleotide (reduced)

Oligonucleotide: short nucleic acid polymers, made up of a strand of DNA or RNA

PBS: Phosphate-Buffered Saline, isotonic salt solution that can mimick the body's pH and ion concentraction

PCR: Polymerase Chain Reaction

pK_a: a measure of a molecule's acidity, representing the negative logarithm of its acid dissociation constant (K_a), a lower pK_a signifies a stronger acid that readily donates protons, while a higher pK_a indicates a weaker acid

RNA: Ribonucleic Acid

SDS: Sodium Dodecyl Sulfate

SDS-PAGE: Sodium Dodecyl Sulfate-Polyacrylamide Gel Electrophoresis

Spectrophotometer: an instrument that measures how a substance absorbs or transmits light (photons) at specific wavelengths to identify substances, determine concentration and analyze certain chemical properties

STP: Standard Temperature and Pressure

Substrate: the molecule acted on by the enzyme

TGS: Tris-Glycine-SDS

TPA: Tissue Plasminogen Activator

UV: Lightwave length between 180-350 nm

VEGF: Vascular Endothelial Growth Factor

V_{max}: Maximal reaction velocity

V_0: Initial reaction velocity

About the Author

Miguel Cervantes-Cervantes was born in Atotonilco el Alto, Jalisco, Mexico. He completed his B.Sc. in Biology and M.Sc. in Biochemistry at the National Polytechnic Institute, in Mexico City. He obtained a Ph.D. in Biochemistry from Rutgers University, New Brunswick, New Jersey, under the direction of Dr. Carl A. Price. Dr. Cervantes' post-doctoral training included work at the Department of Biological Sciences of Rutgers University-Newark with Dr. Douglas Bush. Dr. Cervantes previously held a faculty position at the City University of New York, where his research focused on plant carotenoid biosynthesis. He currently works at the Department of Biological Sciences, Rutgers University-Newark, where he teaches biochemistry, microbiology, and genetics, among other subjects. Dr. Cervantes has taught the cell and molecular biology core course for biology majors and minors for many years. He has also taught a graduate course in college teaching. He has designed hands-on laboratory courses that emphasize experimental approaches to understanding biological processes. Out of the classroom and the laboratory, he enjoys drawing, painting, and traveling to explore cultures different from his own.

For Product Safety Concerns and Information please contact our
EU representative GPSR@taylorandfrancis.com Taylor & Francis
Verlag GmbH, Kaufingerstraße 24, 80331 München, Germany